£2.49.

TRAMS IN THE NORTH WEST

TRAMS IN THE NORTH WEST

Peter Hesketh

Ian Allan Publishing

First published 1995

ISBN 0 7110 2349 2

All rights reserved. No part of this book may be reproduced or transmitted in any form or by any means, electronic or mechanical, including photocopying, recording or by any information storage and retrieval system, without permission from the Publisher in writing.

© Ian Allan Ltd 1995

Published by Ian Allan Publishing an imprint of
Ian Allan Ltd, Terminal House, Station Approach,
Shepperton, Surrey TW17 8AS
Printed by Ian Allan Printing Ltd,
Coombelands House, Coombelands Lane,
Addlestone, Weybridge, Surrey KT15 1HY.

CONTENTS

Preface	7
1. Setting the Scene	9
2. Cumbria	13
Carlisle	14
Barrow-in-Furness	19
3. North Lancashire	27
Morecambe	28
Lancaster	34
4. The Fylde Coast	37
Blackpool	39
Lytham St Annes	55
5. Preston and Dick, Kerr	63
6. East Lancashire and Rossendale	77
Accrington	79
Blackburn	86
Darwen	93
Haslingden	97
Rawtenstall	98
7. End of the Chain	101
Burnley	102
Colne	109
Nelson	114
BCN Joint Transport	118
Fleet List	120
Appendices	126
Acknowledgements	128
Bibliography	128

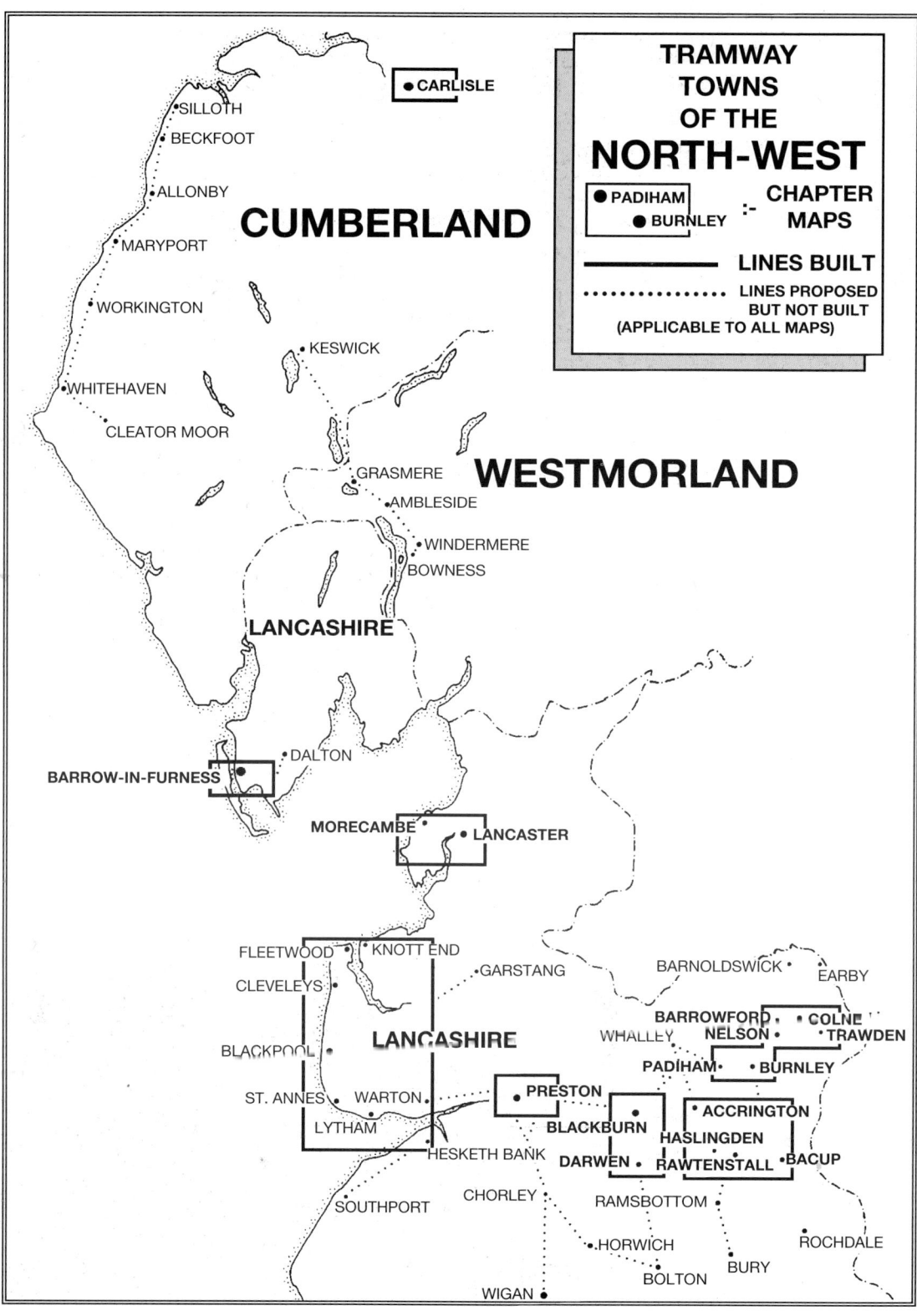

PREFACE

Often beleaguered, historically hapless, the tramcar — now struggling in its slow and painful uncertain rebirth — has, for over a century, seemed to surround itself with foes rather than friends. Even now, in energy conscious times, as the still new Manchester system is showered with sometimes reserved accolades and work progresses on the Sheffield Super Tram, the last remaining traditional system, Blackpool, appears to attract more critics than supporters. Environmentally friendly, of course, the line along the Fylde Coast to Fleetwood continually has to cower from modern-day apostles of the motor car, children and grandchildren of those original advocates of abandonment in favour of the internal combustion engine. The fact that thousands of holidaymakers are whisked daily up and down the system carries no weight with those more mindful of the supposed dangers. 'Keep the traffic moving at all costs; bring in single line working (with red flags?) on the busy street section behind the Metropole Hotel; fence off the promenade reserved track, etc, etc.' All would have potentially financially crippling consequences.

Nothing has changed. In the 1930s, the tramcar rapidly became unloved, even despised. Stuart Pilcher, Manchester's high profile transport manager of the era, is noted to have registered his displeasure in having a tram type named after him, which failed to sit comfortably with his policy of energetic trolley and motorbus introduction, in place of the city's cars.

Leyland Motors, amongst others in the manufacturing field, fuelled the flames, making irresistible offers to corporations as inducements to buy replacement buses. Its aggressive marketing worked. By 1949, only Blackpool remained in the area covered by this book. Thirty years before there had been 15.

If the tramcar in the British Isles has a future, then it certainly has a past; complex, flowered, fascinating and evocative. Let's take a step back in time and look at that past in the North West.

On a cautionary note, though, I must state from the outset that this book is not an in-depth survey; others have already done that in respect of particular areas and operators. This, therefore, is perhaps best described as a detailed overview, the bibliography towards the end hopefully providing ample points of reference for those wishing to study further. The fleet lists are, for similar reasons, in a simplified form, with cars shown 'as delivered'.

PETER HESKETH
Hutton
Preston
September, 1994

In their infinite variety, various forms of traction took to the tramways of the northwest over the years. First came the horse, and, as in the case of Burnley, steam power, depicted here by one of Burnley & District Tramways Co's Falcon locomotives hauling an open-top tramcar. The sulphur fumes were supposedly good for the bronchial tubes!
Neville Lockwood Collection

CLAYTON SECTION
39
ACCRINGTON CORPORATION TRAMWAYS

44
CK 4340

CHAPTER 1

SETTING THE SCENE

Sunrise on a summer's day in 1920. In Barrow-in-Furness, shipyard workers gather on street corners in readiness for their shift, whilst 60 miles or so to the north, across the Lake District mountains, in Carlisle, office charladies scurry along the still unwarmed streets. Simultaneously, hotel waitresses, maids and kitchen porters wait impatiently in Fleetwood, rubbing shoulders with fish dock operatives, before their daily journey down to Blackpool and its still slumbering holidaymakers. In Lancashire's opposite eastern extremity, cobbled streets ring with the resounding cacophony of the clogs of mill girls who pass muted times of day with farm labourers heading away from the drab rows of terraced houses towards the hills and fresh air of the surrounding countryside.

This is an area of stark contrasts. Yet these contrasts encompass one common denominator: the tramcar as a mode of travel. Barrow-in-Furness Corporation to Walney Island, Carlisle & District Tramways to Stanwix, Blackpool & Fleetwood Tramway to Gynn Square and Burnley Corporation Tramways to Brierfield, can be plucked out to epitomise the vista apparent. Within yards, the steep hills of the dark mill-towns become rolling countryside and the deafening din of steelworks gently pales into the peace and solitude of coastal sand dunes, abounding with wildlife. Inevitably, in those times just after World War 1, social contrast was all too prevalent also; affluence and poverty; health and abject misery and disease all rubbed shoulders uncomfortably like those travellers waiting at the tram stop in Fleetwood. This is the Northwest of England in the heyday of the street tramcar, reflecting the infinite variety that was so typical of this mode of transport, born of the Industrial Revolution.

In 1870, Parliament had introduced the Tramways Act, an acknowledgement that what had been regarded a few years earlier as merely a passing craze, was not going to disappear easily. Birkenhead had seen England's first street tramway, laid in 1860 by an American, George Francis Train, and the legislation sought to make future construction easier for promoters, who would have better access to Parliamentary powers. At the same time, though, it also hoped to prevent monopolies; ventures into tramway entrepreneurialism could not be entered into lightly. It was an expensive business, digging up roads and laying rails. Horse omnibus operation was far less complicated and more attractive.

The Act, therefore, imposed considerable restrictions to ensure fair competition. It insisted on such conditions as a requirement for the operator to maintain the roadway between the rails and for 18in on either side of them, and gave opportunities for redress to objectors who felt their livelihoods threatened. By far the most significant part of the Act was the option, after 21 years, for local authorities to purchase systems in their towns. This was a proviso intended to allow councils the ultimate sanction should they be dissatisfied with the manner in which the companies were running things. In some areas, this was a welcomed clause

Above left: By contrast, the electric tram in refinement. Fresh from the Brush Works in Loughborough, Accrington's No 39, acquired by the corporation in 1919. With a capacity of 76, this had longitudinal seats in the lower saloon, with 2 + 1 upstairs, where there were no bulkheads, although the stairheads were enclosed. Demands on the tramcar were now growing, but was the growth sustainable? *Leicestershire Museums — Brush Collection*

Left: The 1930s saw wholesale abandonment of systems, in the case of those in the northwest, succumbing to the onslaught of the motorbus. Preston saw its last cars run in December 1935, and here a Leyland TD2 Titan, one of 10 purchased in 1932 to allow the first closures, passes workmen lifting track on New Hall Lane near the Farringdon Park terminus. The bus has bodywork by English Electric, builder of tramcars in the town, whose story is told later. *Author's Collection*

as, in the case of Barrow-in-Furness, for example, the company's operating standards declined at an alarming rate.

Moving the masses in the towns soon became, more or less, the exclusive preserve of the tramcar, but, perhaps unfortunately, in its infancy it was seen as transport for the working classes. This position only changed on corporation take-overs, heralding the inevitable progress to electrification and the introduction of more hospitable vehicles, acceptable to the middle classes, to replace either the smoke and spark-belching steam-hauled cars or those horse-drawn, slow and laborious cars. Steam tram locomotives were required by law to have most of their motion concealed, to be noiseless and consume their own smoke. They managed the concealment part, but the law was rarely adhered to with regard to the racket and choking fumes. The sulphur, though, was hailed as a cure for many Edwardian ills of the bronchial tubes!

What could have been the salvation of the tramway companies, the Light Railways Act of 1896, came too late to save the established ones from council acquisition. This act more specifically intended to encourage simplified railway branch line construction, by removing the need for Parliamentary powers and relaxing conditions governing signalling etc — as was the case with the Whittingham Hospital Tramway, between Preston and Longridge — was often used, however, by tramway operators to extend their systems, or to enable the starting of new schemes. (Separate legislation was still needed to facilitate such proposals as electrification and full council take-overs.) In Colne, the act title was actually adopted by a new company working to hitherto unserved parts of the town and its outskirts, the Colne & Trawden Light Railway, a private venture, whilst neighbouring Nelson built its system as a municipal-owned operation, as did Lancaster; all of them comparative latecomers to the scene.

It would be a gross understatement to say that opportunities were lost. Exciting proposals at the time could have led to a massive integrated tramway system covering most of the then county of Lancashire. In 1903, the Preston & Blackburn Tramways Co had been formed to build and operate a line between those towns. It came to naught, as the legislation foundered at the Commons committee stage, its death attributable, certainly, to the 10 miles or so of sparsely populated rural area on its proposed route. If it had come about, through running from Bacup, over 30 miles to the east of Preston, would have been possible. There was further potential for a link between the Fylde Coast systems and Preston, but an even shorter-lived company, the Preston and Lytham Tramways & Tramroad, also suffered from the problem of likely rural passenger numbers on its intended 12-mile journey through the Fylde villages. However, meeting up with the densely populated southern part of the county, with its intensive route systems, came closer with the establishment of the Preston, Chorley and Horwich Tramway Co, also in 1903, which proposed three routes, enabling Bolton and Wigan to be reached and to serve the rapidly expanding town of Chorley. In turn, South Manchester could have been reached from Fleetwood, giving a north to south through-running potential of some 70 miles. Gauge rationalisation would have been necessary, though, as Preston had eventually opted for the standard, 4ft 8½in, in line with the Fylde Coast systems, as opposed to the more usual 4ft of the rest of those in the area. However, where through running was possible, in practice the opportunity was not always taken, as will be seen later. There were several other examples of schemes that failed to materialise, given up at varying degrees of progression. Some were obviously doomed to failure, others were exciting prospects, perhaps too advanced for the time. These will be revealed as the story unfolds.

Such was the development; the decline soon followed. In 1931 the Royal Commission on Transport pronounced the death sentence on the trams, in response to gathering antagonism towards them. In brief, they caused congestion and were a danger. The Commission proposed that no new tramways should be constructed and that existing ones should be abandoned as soon as possible. Already, Lancaster and Haslingden had been converted and Accrington followed that very year. By the outbreak of hostilities in 1939, only Blackburn, Darwen and Blackpool remained, the stay of execution on the former pair only to be lifted soon after the war. Trolleybuses were considered and tried by some, but despite powers being obtained to run them, it was the ubiquitous motorbus which swept on to the streets with a vengeance.

During the period of tramway history, characters and personalities rose to prominence. The Lancashire & Cheshire Tramways Authorities Council, on which most corporations had representation, provided a platform for the forceful and often indomitable managers sent along with their committee chairmen to fight their particular tramway corner.

Although Manchester's Stuart Pilcher often dominated this committee (and, indeed, used it successfully as a vehicle for his own advancement) others were equally prominent. Burnley's Henry Mozley was one, at the helm from company days in

The attraction of the motorbus was diverse in itself and reasons for conversion differed from town to town. Lytham St Annes was plagued by drifting sand on its section between Squires Gate and St Annes. No 41, an English Electric-built balcony car of 1924, has come to grief in the encroaching sand on its way to Lytham past the site of the present Pontins holiday camp. It is easy to understand why the bus was to be deemed more practical for this operator. *Roy Marshall Collection*

1882, to his retiring from the corporation successor in 1930. He persevered with the troublesome steam tram and got a degree of efficiency from it; he patented a right-hand spiral staircase and designed the famous 'Burnley Bogie'. Another was Harry Clayton, for 23 years manager at Preston and responsible for that operator's imaginative programme of tramcar building, rebuilding and adapting, a tradition perpetuated at the Deepdale depot to this day. 'Blackpool is not a work-a-day town — why should it have work-a-day trams?' was Yorkshireman Walter Luff's early retort as he embarked on the system's rejuvenation. His streamlined cars survive, plying up and down the Fylde Coast, a lasting epitaph to him and his fellow managers of the golden age.

To complete this scene-setting, the area of study and how it came about must be explained, as this northwest region of the country stretches, in reality, beyond the bounds of what is, in effect, for the purposes here, the old counties of Cumberland and Westmorland and about half of what was traditionally known as Lancashire. Local Government reorganisation in 1974 swept away the two northern counties and imposed Cumbria upon Lakeland and Furness folk, the latter being previously in Lancashire. Likewise, that county was decimated overnight, losing the Liverpool suburbs, St Helens, Southport, and the rich arable farmland between them to Merseyside, whilst such as Wigan and Bolton and the towns surrounding Manchester (Salford, Ashton-under-Lyne, Oldham and Rochdale, etc) passed to the county of Greater Manchester. Further Lancashire losses were encountered to the south, where Warrington and Widnes were swallowed up by Cheshire.

After the upheaval, left remaining were the two new shire authorities of Cumbria and Lancashire, with the former tramway towns of Carlisle, Barrow, Lancaster, Morecambe, Blackpool, Lytham St Annes, Preston, Blackburn, Darwen, Haslingden, Accrington, Rawtenstall, Burnley, Nelson and Colne.

The consequences of the Tramways Act, as expected, proved complicated, with pre-corporation companies, in some instances, operating across boundaries. Initially, the author had intended to present an 'A' (Accrington) to 'R' (Rawtenstall) chronicle of the towns, but research soon showed a need to group, as some were inextricably linked. Of course, there had to be one which would not fit in! Preston was isolated, tramway-wise, and, although bordering the Fylde Plain, has little affinity transport-wise to most of it. Therefore, it stands alone, save for the fact that a small, private system within the town is tagged on. That is the test track for Dick, Kerr, or the United Electric Car Co, to give it its full title, the renowned tramway builder, whose history and products are incorporated into the Preston chapter.

Needless to say, the centre-pins of all these diverse systems were the tramcars themselves, diverse not just in makes and types, but also in their sources of power. Often horse and steam traction were utilised in company days before electrification, and one employed from its outset (although briefly) the conduit electric system, whilst another used petrol. Another most unusual mode of propulsion to be found was gas.

All in all, the tramway story in the northwest abounds with individuality, but, as mentioned earlier, it necessarily follows that growth and development tend to give way to decline. Sadly, perhaps, this is an integral part of the tale. It offers no succour to the enthusiast of the tramcar, but is the final link in the chain. Unless, that is, the tram is reborn.

Top: Although poised for abandonment in the late 1930s, Blackburn's trams survived the war, proving invaluable in moving the masses of workers during those dark days. No 66 was one of only eight of 40 cars, new in 1901, not eventually to receive top covers. The Blackburn system closed in 1949, and in this postwar view it shares the depot with, on the right, one of the 30 Crossley-bodied Guys or Leylands bought to replace it and its cohorts. The bus on the left is one of the pair of Pickering-bodied utility Guys which went to Sunderland in 1947 and is presumed to have escaped being photographed in Blackburn. *A. D. Packer Collection*

Above: Walter Luff's legacy on the Fylde Coast, in the shape of balloon double-decker No 710, which started life as No 247 in October 1934. Built by English Electric, its streamlined body, whilst being refurbished over the years, has lost little of its original lines. Even the livery in this July 1994 view is a reversion to its first, but its trolley pole has been replaced by a pantograph. The location is Fleetwood Ferry, but more about these fine cars in due course. *Author*

CHAPTER 2

CUMBRIA

Had the 19th century visionaries had their way, this vast area would have been riddled with a not unreasonable number of railways and tramway systems. Of the latter, whilst one — a proposed and rejected plan in 1899 to construct an electrified line from Bowness Pier, up to Windermere station and then on to Grasmere — was probably doomed from the outset, despite an abortive attempt to revive it in 1921 as a petrol-electric-powered tramway further northwards over Dunmail Raise to Keswick, another did have merit. Parliamentary powers were obtained in 1901 for a 31-mile standard gauge tramway along the Cumberland coast. As the West Cumberland Electric Railway, it would have run from Silloth in the north, down to Whitehaven and then inland to Cleator Moor, passing through such places as Beckfoot, Allonby, Maryport and Workington. As it was, only two projects, those of Carlisle and Barrow-in-Furness, evolved far enough to see cars on their metals.

Despite early proposals for a horse tramway, Carlisle moved straight into the electric era in 1900. Fittingly, the first photograph in this section shows the 3ft 6in gauge track being laid in the city centre at the junction of Lowther Street with Warwick Road and The Crescent. German rail, at 83lb to the yard, was used. More complex trackwork was assembled at Dick, Kerr's Kilmarnock works and relaid *in situ*. In later years (post-1912) the layout was simplified and singled. *Carlisle Library*

CARLISLE

If this isolated 3ft 6in gauge system has a claim to fame, it has to be the circumstances of its end, rather than its beginning. It epitomised the anti-tramway feeling growing in the 1920s and suggests a strong leaning by officialdom towards the motorbus. However, first things first. In 1880, or thereabouts, there had been talk in the city of proposals for a horse tramway. Talk it was only to be and Parliamentary powers were never sought, but nearly 20 years later, the Carlisle Tramways Order, 1898, gave a Birmingham syndicate the go-ahead to construct and run an electric tramway, under the auspices of the Manchester Traction Co Ltd, registered the year before.

The order authorised no fewer than 27 separate tramways totalling a mere 5.5 route miles or 7.75 miles of track. As will be seen, this was usual practice, splitting lines into sections and treating short lengths, loops and crossovers, etc, as tramways. In reality, six lines were approved: to Newtown, Stanwix, along Warwick Road to Petteril Bridge, London Road, Boundary Road and Denton Holme, all radiating from the city centre.

Carlisle Corporation gave its consent on the proviso that all electricity was to be purchased from the council's power station and, under the Tramways Act, it would have the usual option to purchase the system after 21 years, and at seven year periods after that. The operating company emerged as the City of Carlisle Electric Tramways Co Ltd, with its registered office at 23 London Road. Again, under the Tramways Act, it would be required to maintain the carriageway, as explained in the first chapter.

The city, whilst being far from the largest in the land, had its fair share of narrow streets (hence the choice of gauge). In one instance 14ft 6in was all that was available for the tramway engineers, the Birmingham firms of Dickinsons and Pritchard Green. They contracted Dick, Kerr to construct the trackwork and overhead, and to supply the cars — three closed single-deckers and 12 open-top double-deckers — built at its Electric Railway & Tramway Carriage Works at Preston. The livery was chocolate and cream.

It took only from September 1899 to June the following year to complete the work, using 83lb/yd rail of German manufacture. As agreed, the power, at 500V dc, was fed along trackside conduit from the power station in James Street, which required

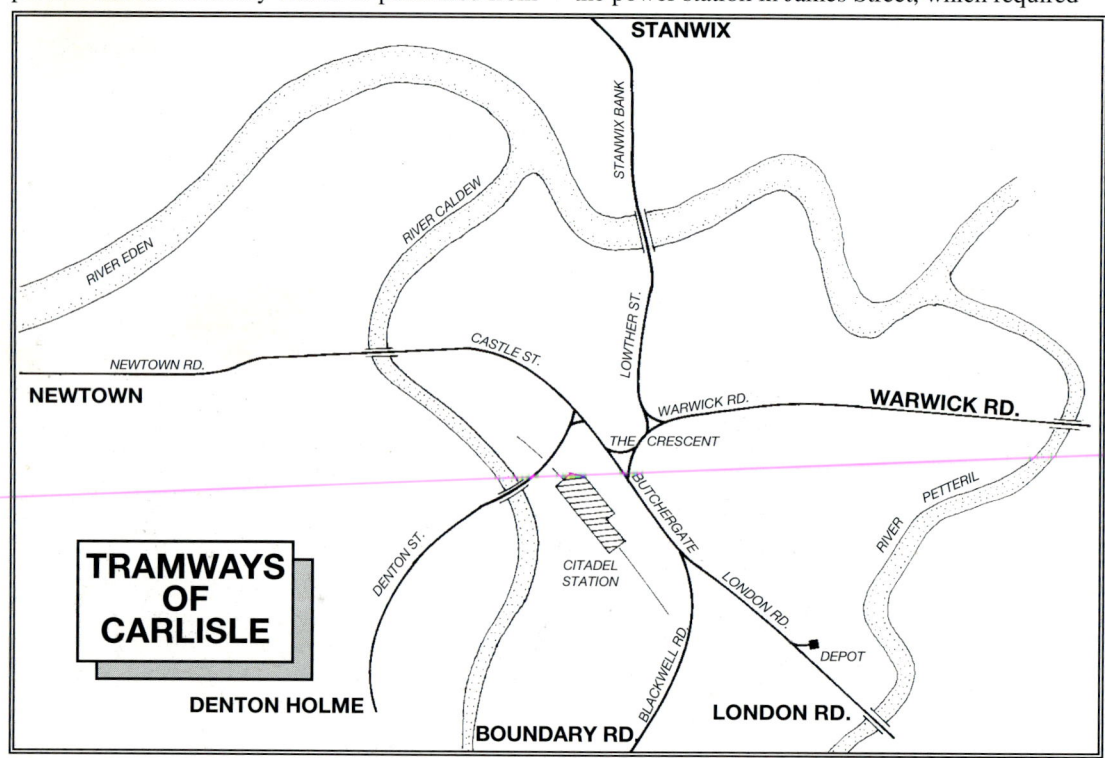

Above: Early days in Carlisle, with one of the first batch of open top ER&TCW cars, delivered for the opening of the system, in English Street, by the Earl of Lonsdale statue. Note the position of the large headlamp. These cars seated 45 and were mounted on Brill trucks. The fleet was owned and operated by the City of Carlisle Electric Tramways Co, the council never securing ownership of it. *Roy Marshall Collection*

Below: Also delivered in readiness for the opening were three single deckers, again built by ER&TCW. One of these, No 2, seen here, was converted for one-man working in 1907 and was destined to remain the only vestibuled car in the fleet. When photographed at Viaduct Corner in 1908, it was obviously being crew-worked on the Denton Holme route. *National Tramway Museum*

two additional Lancashire boilers to cope with the extra load. The finishing (and vital) touch was the company's impressive depot in Lindisfarne Road, near to the terminus of the London Road line. Most of the system was single track with passing loops, except for along Lowther Street to a point just north of Eden Bridge, and a couple of short sections on the Newtown route. All in all, 7.25 route miles were laid, giving a total track length of 8.5 at the opening on 30 June 1900, heralding a great future.

Ten years later, with falling patronage and soaring overheads, especially in respect of the purchase of electricity (no pun intended!) the company was in dire straits, not having paid a dividend to shareholders since 1906. The whole system was in imminent need of replacement, and so, in May, the company sold out to the established tramway operators, Balfour Beatty, the transfer taking place in November 1911. Negotiations with the corporation were successful in bringing about a reduction in the price of current, and the company then felt in a position to start renewing what had rapidly become a dilapidated system. The poles needed raising and the overhead replacing, as did the still quite youthful cars, 12 UEC-built vehicles being delivered in readiness for the opening of the new operations on 9 December 1912. The change of ownership was also marked by the introduction of a fresh livery, dark green and cream.

No sooner had the accounts swung into a small credit (£33) in 1913, than the outbreak of World War 1 in 1914 set into motion what was to become a familiar chain of events. Lack of maintenance and the calling-up of drivers, conductors and craftsmen, gradually began to have an adverse effect, from which the company never really recovered.

Carlisle began to expand after the war, as the population moved out into the new suburbs, and the corporation pressed the company to extend its lines to cater for them. The request fell on deaf ears. In any case, the parent company was not awash with money; what it had was needed to keep its existing systems running. Motorbus operators were quick to fill the void and by the mid-1920s nearly 40 were jostling for passengers in this part of Cumberland. The tramway company had also embarked on such a venture by running buses to Annan, Dumfries, Brampton and Langholm, under the Percival's Motor Bus Services name. By this time (1926) it had appropriately changed its own title to 'Transport' in place of 'Tramways'.

The city council was unhappy with the continuing state of affairs, and in 1923 had formed a committee to examine the possibility of running buses into the suburbs themselves. The tramcars were becoming unpopular and were seen as noisy, unreliable and a cause of congestion in the narrow streets, which were having to cater for the increasing volume of traffic.

As the decade changed, an aggressor with a considerable success rate behind it (which is to raise its head more than once in this book) appeared on the scene. Ribble Motor Services, based in Preston, but expanding northwards, got wind of discussions being held between the company and the corporation in 1930, aimed at developing the tramway. Ribble made an offer for it, with a view to replacement by buses, but the corporation was not in favour, as it wished to establish its own municipal undertaking. On 5 March 1931, the council decided to acquire the tramway and Percival's, its offer of £32,500 being acceptable to the company. The Ministry of Transport agreed to the take-over, but the Traffic Commissioners did not, and refused the corporation a licence, no doubt from the case put forward by the railway companies and the established post-Road Traffic Act bus operators.

An appeal was considered by the council, but, after deliberation, it was decided against, and instead it invited the local concerns, Cumberland, Caledonian (later to become Western SMT), United and Ribble (who had stealthily gained ground by buying out a number of smaller companies) to enter into a co-ordinated agreement. On 13 October 1931, the Traffic Commissioners approved the proposals, and the ratepayers of the city were, in reality, saved from the financial burden, in not too easy times, which the purchase of the company and setting up of bus operations would have created.

Although scheduled to finish in April 1932, the last tram actually ran, with due pomp and ceremony, on 21 November 1931, to enable Ribble to obtain additional licences, which along with United's, were finally granted on 29 December.

Below left: An imposing depot was built in Lindisfarne Road, off London Road. This view, taken at about the time of the opening, shows one of the first cars, without a fleet number. *National Tramway Museum*

Right: This 1900 view shows the double trackwork at the top of Warwick Road, outside the Citadel Temperance Hotel. As all railway enthusiasts know, the station in the city is still known as Carlisle Citadel. *Carlisle Library*

Right: No 14 at the Newtown terminus, *circa* 1905, by which time the large headlamp had been removed and replaced by a more conventional one on the front dash panel. *Carlisle Library*

Below: Lowther Street, looking towards The Crescent, with car No 12 running in from Stanwix. *Carlisle Library*

Left: Utter and complete dilapidation in Barrow! A former Barrow-in-Furness Tramways Co steam tram towards the end of its life in 1903, some four years after the British Electric Traction Company had purchased the system with the intention of electrifying it. *National Tramway Museum*

Below: Five of the first cars were single-deck combinations, also by Brush. Is the crowd more interested in the new car or the photographer? *National Tramway Museum*

Below: A Barrow-in-Furness Tramways Co ticket, showing all routes. *National Tramway Museum*

Right: A new broom! No 5 in pristine condition. This was one of the 12 cars bought by BET in 1903 on the electrification of the system. In common with most of the fleet over the years, it was built by Brush of Loughborough. It seated 46, and although its trucks were also of Brush manufacture, the motors and controllers were Dick, Kerr. A livery of maroon and cream was applied. *Roy Marshall Collection*

to look at ways of implementing the Act. Its first task was to travel the tramway globe, looking at other towns' systems, as by then a fair number had already joined the rush to lay rails and outdo their neighbours in the degree of civic pride gleanable from the new phenomenon. The northeast was visited, with inspections of Tynemouth, Middlesbrough, Stockton, Sunderland and South Shields, as was the electric railway near the Giant's Causeway in Ireland. Back closer to home, fact-finding forays were undertaken to Bolton, Bury, Birkenhead, Southport and Preston. As will be seen later, Preston had just opened its enlarged horse tramcar system.

On return, the deputations reported, and recommended the adoption of steam traction and a gauge of 4ft, reckoning this would save £200 per mile of single line, when compared with the standard 4ft 8½in. A width restriction of 6ft 6in on tramcars was also advised. These all being accepted by the committee, offers were sought from promoters, and it was agreed that the Barrow-in-Furness Tramways Co Ltd, set up by a Manchester tramway promoter, Mr Vawser, should construct and eventually operate the tramway, under the provisions of the Tramway Act, 1870.

The tramway carried its first passengers on 11 July 1885. It had taken just over a year to lay the track for the four routes decided upon, which were: (a) Town Hall to the Abbey, along Duke Street to Ramsden Square, where it turned northeastwards along Abbey Road; (b) a short route from Ramsden Square to the steelworks; (c) Town Hall to Roose station, along Salthouse Road and Roose Road; and (d) Town Hall to Ramsden Dock, passing over the High Level Bridge, where the track was doubled, and down to the forecourt of the Dock railway station.

However, the bridge had, at this stage, not been finished, and it was to be another 12 months before cars were able to reach the terminus. The Abbey line was the longest, necessitating the provision of a water-tank, set into the roadway at the outer end. The layout at Ramsden Square, where the steelworks route diverged, incorporated a reversing triangle into its passing loop. In total, the route length was 5.5 miles, loops taking the track length to six.

Although the work was closely supervised by the Borough Engineer, it was contracted to a Mr Fell. The depot was built adjacent to Vulcan Foundry in Salthouse Road, on the Roose route, to house the eight tramway locomotives. These were constructed by the Leeds firm of Kitson and were fully enclosed, being the requirement, and fitted with the also mandatory roof-mounted condensers, supposedly for them to consume their own smoke. Eight trailers were built by Brush of Loughborough, seating 58 — 28 inside, the remainder on top to inhale the sulphur fumes from the belching locomotive. The formations were painted dark red with white lower panels.

It took some time before local people adapted to their new (and, in reality, first) mode of easily accessible transport. Fatalities were not uncommon, children, sadly, being particularly vulnerable, and three were killed in the early years. In fact, the whole steam-tram operation was dogged with bad luck and attracted ill-feeling towards it, despite all its advantages. No doubt relevant to its poor accident record, was the need for the Board of Trade to admonish the company over dangerous driving by some of its motormen, asking them to ensure they exercised more caution, especially at passing places.

These problems apart, Board of Trade permission was granted in 1893, allowing the use of steam traction until 1900. The company saw an opportunity for expansion in this seven year period and its secretary asked the council to seek Parliamentary powers to extend its routes. An extension off the Abbey route at Claye's Mansion to the Water Trough at the Abbey and then to Dalton-in-Furness, some five miles farther on, was proposed. This would utilise the new road under construction. Also suggested was an off-shoot of the Ramsden Dock line to Walney Ferry. The council did not accede to the request, although Walney Ferry did see a service in later years. Dalton was reached, eventually, but by corporation buses, in 1915.

The latter years of the 19th century witnessed a deteriorating state of affairs, not just in the condition of the tramway, but also between the company and the corporation, who repeatedly asked for repairs to be undertaken on the trackwork, and for the cars to be maintained to an acceptable standard. It was reported that, for example, the locomotives' condensers were missing, either having been removed as defective, or fallen off from the vibration. Management of the company appears to have been cavalier, and timetables became less meaningful as time went by. A particularly low point came in January 1897, when the corporation threatened the company with legal action unless matters improved. Work was put in hand and the threat was not carried through.

Final ignominy for the company came, perhaps not surprisingly, in 1898, when it went into liquidation. The Tramways Act option gave the corporation the opportunity to acquire the operations and this was considered at the council meeting on 18 November; £22,750 was the asking price, but the council declined. Steam power could not continue, as apart from the locomotives being worn out and dangerous, electricity was now the accepted and, indeed, most efficient source of power. There was no place in the 20th century's streets for tramways of the horse, iron or otherwise. Thus, after giving an assurance that improvements would be made immediately, the British Electric Traction Co, another operator of systems across the country, took over on 23 December 1899, with the intention to electrify. In the meantime, the steam trams continued, but, in spite of the efforts by the new owners, the system became so precariously worn, that the Board of Trade could issue its certificates for steam operation only on a monthly basis. The locomotives, leaking steam from every conceivable joint, time and time again failed to get up the town's not all that steep hills, having to reverse and take more than one run at them! Derailments became more frequent. The whole system was literally falling apart at the seams. To top it all, passengers were getting fed up with the state of the conductors, whose filthy conditions rivalled that of locomotives and cars. Things were far from helped by a fire at the depot on the evening of 27 June 1902, which put the system out of action for a time.

The following month, the Chairman of the Tramways Committee and the Borough Engineer went off to look at electric tramcars elsewhere, including Lancaster, and also visited the Brush Works at Loughborough. As expected, sooner rather than later, the tramway ground to a halt in the summer and BET, under the Parliamentary powers obtained by the corporation, began work on electrifying the system. Also obtained had been an agreement with the Furness Railway to lay tracks to Walney Island over the proposed bridge replacing the ferry. They owned the lines alongside Ferry Road and also the steam ferry.

Electrification of the system was authorised under the Barrow-in-Furness Tramways Order of 1903. This was undertaken, along with the complete replacement of the permanent way (again to 4ft)

Below: The decrepit state to which the steam fleet descended was, sadly, mirrored by the BET-owned electric cars, as depicted here by a shabby and sagging No 20, one of the quartet of Brush 96-seaters, delivered in 1911 to cope with the rapidly developing shipyards and the excursion traffic to and from the steamers plying between Ramsden Dock and Fleetwood, Blackpool and the Isle of Man. This particular car, along with No 19, was to lose its top deck in 1928 when it was also vestibuled. It is seen here on its usual haunt near the High Level Bridge. *Roy Marshall Collection*

Above: Now in corporation ownership is former BET car No 5. Barrow Corporation took over the ailing company in 1920, and 12 months later started painting the fleet in a new olive green and cream livery. It is seen at the Abbey terminus. *National Tramway Museum*

Below: By now vestibuled, 96-seater, No 17, was, like sister No 18, destined to keep its top deck. Its destination blind shows the short working to the Tea House on the Ramsden Dock line. *National Tramway Museum*

sub-contracted by Brush to Griffiths of London, rails being supplied from Belgium, and an extension to the depot, in time for the official openings, firstly of the Abbey and Roose lines on 6 February 1904. The Ramsden Dock and Walney Ferry routes followed in June and October. Total route mileage was 5.5 miles, with track length 6.87, the first figure being the same as in steam days, this being attributable to the early abandonment of the steelworks line in the 1890s.

Twelve tramcars were required, seven double and five single-deck, the latter being needed to negotiate the low railway bridge at Salthouse, on the Roose service. They were painted in BET's standard maroon and cream, and were augmented the following January with the arrival of two small 'demi-cars' for use at off-peak times on the through Ramsden Dock to Roose service.

These cars were claimed to halve electricity consumption and reduce overall costs. Two additional double-deckers appeared towards the end of the same year.

Perhaps remembering the infamous recklessness of the old steam tram drivers, and, of course, mindful to maintain the profit recorded in its first year, Mr Blaydon, the manager, in 1905 devised a method to economise on electricity, based on the elimination of careless driving, often stemming from drivers racing to termini, to, it was rumoured, study racing form or have a cigarette! He drove a tramcar over the system, and, on passing each pole, noted the position of the controller. Thus, the poles could be marked 'off', 'series' or 'parallel', and drivers were instructed to adhere to them. It must have worked, as consumption fell from 1.03 units per car mile to 0.96.

Building of the opening bridge to Walney Island started in 1905, much to the chagrin of the Furness Railway, who, of course, owned the ferry. The tram tracks in Ferry Road were relaid on the renamed New Bridge Road towards the bridge. Control of traffic over it (which included a railway line between the proposed double tramtracks) was to be by standard Furness Railway signals. It was opened on 30 July 1908, and the trackwork on it was ready for inspection by the Board of Trade on 8 June the ensuing year, by which time work was progressing on an extension to Biggar Bank on the island. Constructed by the corporation, it cost £7,000 and was commissioned in August 1911. Tramcars passing over the bridge were subject to a toll, and passengers purchased an extra ha'penny ticket to cover this. The corporation were also obliged to pay 5d (2p) per car mile on the extension to BET. On the car front, four mammoth 96-seaters were bought in 1910 from Brush to handle the ever-increasing shipyard traffic, and were restricted to workers' services and those meeting steamers from Blackpool, Fleetwood and the Isle of Man.

On the outbreak of World War 1 in 1914, Barrow's tramways, besides having to bear additional burdens in moving increased numbers of workers, suffered from restrictions imposed by military activity on the Biggar Bank route for the duration of hostilities, which cost the corporation £400 a year from 1916 as a rebate to the company for the closure of the service. Shipyard workers on a weekend basis and women conductors were recruited to help maintain services. In 1917, they saw such a rise in demand caused by a sudden surge in munitions manufacture, that two trailer cars had to be acquired to further boost the fleet which had been increased by the arrival of two cars from the Potteries fleet two years before.

By 1918, the system had slipped into a perilous state, and the corporation were told that the company could not continue running the number of cars it increasingly had been doing. Return to peacetime conditions did not help; reductions in workers' hours and other spiralling costs were pushing it towards an alarming mirror-image of the dilapidated steam tramway it had so triumphantly replaced only 15 years before. On 1 January 1920, the corporation invoked the provisions of the Tramways Act and the Barrow-in-Furness Tramways Orders and purchased the company. It was reckoned that £120,000 would be needed over two years to bring the tramways up to scratch and the corporation was authorised to borrow £107,750.

On take-over, the corporation found themselves owners of 25 cars, a horse-drawn tower wagon and a water car. Six of the cars were beyond repair, and consequently 12 single-deckers were ordered from Brush. As an urgent stop-gap, four Californian-type cars arrived from Southport in June and, in September, six ex-Sheffield cars were bought, which at first ran in that city's colours. A new livery of olive green and cream was applied to all the cars from 1921.

Various efforts by the Tramways Department failed to make the system pay, and in 1922 an official inquiry by the Tramways Committee was demanded, but came to nothing, despite the trams costing the ratepayers 8d (3p) in the pound. This was cruelly ironic, as it had the lowest operating costs per car mile in the country. Yet, in November 1926, to see if costs could be reduced further, a bus service was introduced over part of the Ramsden Dock route at weekends. Costs were reduced, and for the rest of the winter, the cars were used only at peak times, buses working the remainder as far as the Tea House — the thin end of the abandonment wedge.

Tram scrapping was considered in earnest in 1930, a number of cars already having been broken up, and the manager reported on his inspection of Wolverhampton trolleybuses. Two years later, after the rejection of a proposal to spend in excess of £127,000 to modernise the tramway, incorporating modern rolling stock or the conversion to trolleybuses (deemed impractical however, because of the two opening bridges), it was agreed that 18 Crossley double-deck buses be purchased.

Final tram day was 5 April 1932, cold, wet and windy, which attracted few well-wishers. In his detailed history of the system, Ian Cormack quotes from a speech made by Thomas Lord, Barrow's General Manager, 26 years later in 1958 — 'There is also the possibility of returning to electricity as our motive power by the increased application of nuclear energy for industrial uses'. Barrow Borough Transport succumbed to Ribble in the post-deregulation upheaval of the late 1980s, as Carlisle had done many years before; the misfortunes of the town's tramcars finally echoed by the all-conquering buses.

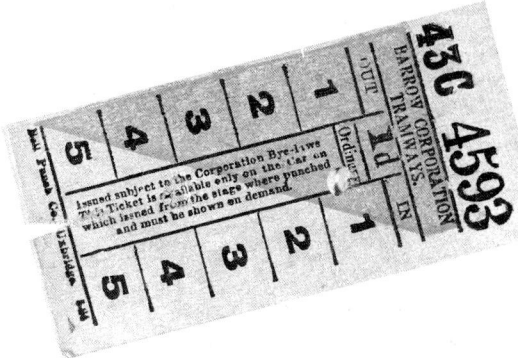

Below: The earlier company tickets showing all destinations, gave way to the fare stage variety of the corporation. *National Tramway Museum*

Below: The final delivery of cars to Barrow was a batch of 12 single-deckers, inevitably from Brush, in 1921. In this builder's photograph, No 41 clearly shows off its Peckham P22 trucks on which they were mounted. These were the only new cars acquired by the corporation after the take-over of the system. *Leicestershire Museums — Brush Collection*

Top: Plying between Lancaster and Morecambe, the Lancaster and District Tramways Co operated a standard gauge line opened in 1890. It used 14 horse-drawn double-deck cars, built by the local Lancaster Carriage and Wagon Company, seating 40. This is the Lancaster terminus at Stonewell, with car No 11. It would have left for Morecambe traversing the street straight in front of it, and arrived by that to the right in front of the church. The line in the foreground is a short stub. *National Tramway Museum*

Above: Also in Lancaster is this peculiar car of the Lancaster company. It is, in fact, one of the double-deckers converted to single-deck, by lowering the top deck to the lower waist level, creating a 'raised platform' car. The line closed in 1921. *National Tramway Museum*

CHAPTER 3

NORTH LANCASHIRE

Cruel things have been said from time to time about Morecambe, usually by latter-day music-hall comedians, and in recent times what would be considered by many to be a far more cruel blow was struck against Lancaster City Transport, victim of the post-bus deregulation 'situation'. Bus operation in the district is now firmly in the hands of the private sector's Stagecoach Holdings' subsidiary, Ribble, a company which has, since its birth in an earlier life in 1919, been prominent in Lancaster and Morecambe. In 1927, the aggressive acquisition policy by the youthful Ribble saw the purchase of a firm called Fahy's of Morecambe, (within the Lancashire and Westmorland Motor Services take-over) who operated a service between Lancaster and Morecambe. This had come about by their merger in 1914 with the Lancaster & District Tramway Co, a horse tram operator. Buses and trams were run in conjunction until 1921, when the tramway was finally abandoned. Before the activities of Lancaster and Morecambe themselves can be examined, it is, however, necessary to recount the Lancaster & District story, which displays all the irony witnessed more recently, especially as it was not to be until 1979 and five years after the amalgamation of the municipal undertakings, that local authority running between the two former boroughs came about.

In 1888, despite there already being a railway link between Lancaster and Morecambe, the Lancaster & District Tramway Co was formed, hopefully to exploit the need to serve a population of some 46,000 and compete with the 130 or so carriages reputed to have worked the road. The terminus at the Morecambe end was to be on Market Street, near to the Royalty Theatre, and although the Lancaster end was to terminate at the Stonewell, giving a total length of four miles, two furlongs, four chains, there were plans to extend, through the city centre and southwards to Scotforth, another one mile further on. From the outset, the company stated its intention to electrify the standard gauge route, but to use horses in the interim.

The contract for the Morecambe-Stonewell section went to a Mr Goldsworthy of St Helens, who started work in January 1890, with considerable alacrity, the whole length being ready for inspection by the beginning of August. Although problems were encountered on the day of the inspection, 2 August, when the first corner to be negotiated, on Cable Street, caused the car to derail, the line was deemed safe and a licence for six months was issued.

Apart from a short double line section in Torrisholme Village, it was all single, with passing loops, and at the Lancaster end a triangle was formed along Cable Street, North Road and Chapel Street. As would be expected, the line was more or less level, except for the part over Cross Hill which required the attachment of a 'trace' horse to assist. The original fleet of 14 cars was housed in a depot in Lancaster Road, Morecambe, which, with the progression of time, eventually became a Ribble bus garage. The offices were at first situated at Stonewell, but later moved to St Leonard's Gate. As much of the company's trade was seasonal, the majority of its horses were sold to other operators in the autumn; in 1895, for example, 85 of the total of 97 were disposed of.

Electrification never came to Lancaster & District. It was unable to compete with the railway, which took only 15min compared with 35 on the trams, at a charge of 3d in 1911, as opposed to the tramway's 4d. Three years before, in 1908, in an effort to balance the books, it tried to substitute some of its journeys with a motorbus. It lasted four months, but made no significant impact on its finances. However, as already mentioned, buses reappeared as supplements to the trams in 1914, and thereafter the tramway disappeared.

MORECAMBE

The Morecambe tramway story is far from straightforward, as with many other parts of the region. Originally the village of Poulton, the resort of Morecambe as it is today was born of a three-way marriage between that and the outlying villages of Torrisholme and Bare, nurtured by the arrival of the Midland Railway in 1848. The deluge of visitors was met at the station by myriad wagonettes and carriages, which were soon unable to cope. Thus, in 1886, the Morecambe Tramways Act, promoted by local businessmen who had established the Morecambe Tramways Co the previous year, authorised two lines. The first, from the recently completed Central Pier, along the Promenade to the Battery Hotel, was one mile long and opened on 3 June 1887. Other than a crossing over the railway yard's 24 tracks by the Jetty, its construction had proved uncomplicated and, except for a double line section at the pier end, it was single track with two passing loops.

The trams were to be horse drawn, and for the opening, four cars were acquired from the Lancaster Carriage & Wagon Co, who supplied vehicles to all the local concerns. The cars were an instant success, and at the company's first Annual General Meeting, a profit of £338 was announced; yet, as was to be proved ironic later, the first of many calls for electrification was dismissed out of hand.

The second line, southwards from the Battery Hotel to Strawberry Gardens, commenced operations on 19 May 1888. Two more cars were needed for the extension, which lay within the Borough of Heysham, and passed over a hillock, known as Cross Copp, a few yards on the Heysham side of the depot, (still in use as another of Ribble's garages, taken over from Lancaster City Transport in 1993). This obstacle, like that on the Lancaster & District, required 'trace' horses. In 1890, the double line section at the pier was extended to the railway crossings, and, in the following year, the section from the Battery to a point near Albert Road was, likewise, doubled. Of course, at this time the Lancaster & District line had reached its Morecambe terminus, but there was never a connection between the two companies' tracks, despite them running parallel for a short distance on the Promenade.

The Morecambe Tramways Co saw the return on its investment growing, so in 1892 the Tramways Orders Confirmation Act permitted further expansion in the form of three new tramways.

However, under this Order, only one was constructed. The new line, from Central Pier, northeastwards to East View, was opened on 17 June 1895, giving the company three miles of tramway between there and Strawberry Gardens. What of the other two authorised tramways? By 1897, Morecambe Corporation had started to show an interest in the tramway (and, no doubt, its profitability). The powers granted under the 1892 Act had expired, which prompted the corporation to seek fresh authorisation under the Confirmation No 2 Act, that year, for the building of a tramway from East View to Bare. (The 1892 proposal for another line to Regents Park was never resurrected.) The corporation were to construct the line and lease it to the company for 10 years, under a contract made on 16 May 1896, when the company agreed to pay £150 a year. It was ready for opening on 23 March 1898, with four new double-deck cars being bought for it.

During the final decade of the century, a number of attempts were made to buy the tramway, most notable being that by the British Tramway & Light Railway Association in 1898. They offered £33,247 for it, but the company and the corporation held tight. Things settled down as the century turned, but in 1906 another protagonist appeared on the scene in the form of one George Balfour, who proposed the formation of a holding company to acquire the system. Having watched Lancaster & District's fortunes wane, the ratepayers of Morecambe began to voice their opposition to the tramcar, a reaction to the corporation's moves towards a take-over of the company. Councillors had become alarmed at the thought of Balfour's plans and were also anxious to include the tramway in their modernisation of the seafront.

Heedless of the rumblings from their residents, culminating in noisy public meetings, the corporation pressed ahead and sought professional opinion from Chester Tramways, who advised it to exercise its Tramways Act options and take over the line in full and electrify it. Company and corporation, the latter deciding to cast caution to the wind, started negotiations in July 1908, but could not agree on a price; the company wanted £15,592, the corporation would only offer £7,661. On 18 November the approval for the corporation take-over and for the double tracking was received, but there was still deadlock, resulting in the matter going to arbitration in London. On 3 February 1909

Above: A Morecambe Tramways Co horse tram heading along the Marine Promenade. All the fleet is believed to have been built by the Lancaster Carriage & Wagon Co, except for perhaps seven second-hand cars acquired in later years. The double track dates the view as post-1891, as explained in the text. *Author's Collection*

Below: Another picture postcard view taken about the same time near the jetty, which was well used at this time by pleasure steamers. The car carries the intermediate all-green livery, which superseded the initial maroon, teak and white and came before the final maroon and white. *Author's Collection*

Right: The straightforwardness of horse tramway operation reflected in a Morecambe Corporation ticket. *National Tramway Museum*

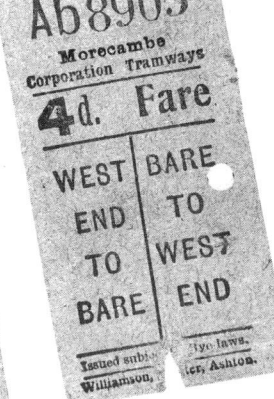

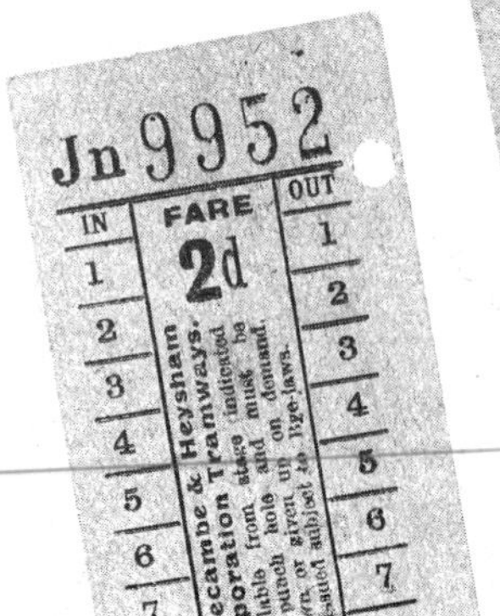

Left: Amalgamation of Morecambe and Heysham councils towards the end of tramway operation is evident in this ticket issued by the new borough's system. *National Tramway Museum*

the arbiters announced their decision — £13,391 — and the council coughed up.

Morecambe Corporation did not get all it wanted, though, as the section from the Battery to Strawberry Gardens was to remain in the hands of the company. The Battery-East View section, with running powers into its part of the depot (now in company territory) were in the deal, which included 12 of the cars. Hand-over was scheduled for 26 July, and came to pass without the civic pomp normally associated with such occasions, the council being more than aware of the ill-feeling locally about its spending of public funds. The first corporation tram ran two days later.

As already approved, the track was doubled during the winter of 1910/11, with a grant of £12,225 from the Local Government Board, but electrification was still being resisted. The sea front and Promenade were now the town's showpieces and unsightly overhead wires were not considered aesthetic. Although the purchase of Edison-Beach battery vehicles was discussed, the horses held sway.

The Morecambe Tramways Co, now a shadow of its former self, soldiered on in private ownership, albeit with only three cars for its one mile route, and a small part of the depot. It was not short of ready cash, though, the sale to the corporation having left it with considerable capital, which enabled it to reinvest in stock. Small it was, but

Left: Victorian Morecambe, with horse tram No 5 waiting for its motive power. In the background is one of the pair of toastracks, built by Lancaster the year before No 5, in 1887. *National Tramway Museum*

enterprising, nevertheless, and in 1912 it took delivery of four trams to replace the horse-drawn vehicles. These were, however, unique and novel in their own right, being self-propelled petrol cars. Their chassis had been built by Leyland Motors, with bodywork by UEC. Leyland engines, naturally, were fitted. These were rated at 55hp. Radiators were mounted at each end.

The cars, all single-deck and three of them enclosed, were immediately acclaimed, but the company considered it wise to replace the 23-year-old 65lb/yd rail with 90lb lengths to withstand the use by them. This necessitated the hiring of a horse bus during the winter of 1911/12, whilst work was undertaken. (In subsequent years one petrol tram was proved to be ample during the off-season.)

All seemed quite rosy for the Morecambe Tramway Co; it had survived World War 1 by converting the closed cars to run on mains gas, with reservoir bags mounted on the roofs, but in 1918 it bought three buses. It had seen several small operators spring up locally and work into Heysham Village, beyond Strawberry Gardens. Competition was, in the main, 'seen off', and the petrol trams ran alongside the company's buses for some six years until 24 October 1924. On that date, the trams were withdrawn and immediately scrapped by the shipbreakers at the Jetty.

That, of course, was not the end of the company. A landmark was reached on 13 December, when it became the major shareholder in a new concern, Heysham & District Motors Ltd, which stated its intention to acquire all the Tramway Co's assets as a going concern, but to sell it when the time was right. The Tramway Co had reached an agreement with Morecambe Corporation in connection with through running between Morecambe and Heysham, and on 17 February 1926, the Tramway Co went into voluntary liquidation, with the 'goodwill' and buses passing to Heysham & District. What was left, which included the depot, was bought by Heysham Corporation. This was only a stop-gap as Heysham & District wanted to sell out completely to the corporation. It was not until 4 May 1929 that it was finally sold, and not to Heysham Corporation, but to the newly-formed Borough of Morecambe & Heysham.

Returning to Morecambe itself, the rift between its supporters and opponents was never healed. In 1918 Parliamentary permission had been obtained to operate motorbuses besides its cars, and two single-deckers had been bought in 1919. As the 1920s progressed, more bus services were introduced, including one to Heysham, with the agreement of the Tramway Co in 1924.

On 6 October 1926, Morecambe, the last horse-tram operator on the mainland, ran its final car. Two aldermen, Hall and Gorton, having shared the driving on the last journey, accepted they had to bow to progress, but, in reality, they were bowing to public pressure, which, on reflection, may not have been so great had the system been electrified, upgraded and seen to be efficient.

Below: The 'trace' horse can just be made out in this view of a Morecambe car ascending the hill at Cross Cop on Heysham Road. This extra horse was needed to assist it up the gradient. *National Tramway Museum*

Above: In July 1909, Morecambe Corporation acquired the Tramways Co, but the section between the Battery and Heysham was not part of the deal, so, in effect, the company survived, albeit in a smaller way. Now not short of funds, it abandoned horse working and invested in four petrol-powered cars, constructed by UEC and incorporating Leyland mechanical units, including a 55hp engine, with a radiator at each end. Awaiting departure from outside the company's offices at the Battery is No 1, new in 1911. *National Tramway Museum*

Below left: On 6 October 1926 Morecambe & Heysham Corporation, owner of the mainland's last horse passenger tramway, ceased operations. That day, car No 8 of 1897, makes one of its last journeys, without ceremony or well-wishers. *National Tramway Museum*

Above: In 1913, the final petrol car appeared, this time as an open-topper, No 4. *National Tramway Museum*

key
1. WILLOW LANE
2. STATION
3. STONEWELL
4. DEPOT
5. SOUTH RD.
6. BOWERHAM RD.
7. QUERNMORE RD.

TRAMWAYS IN LANCASTER AND MORECAMBE

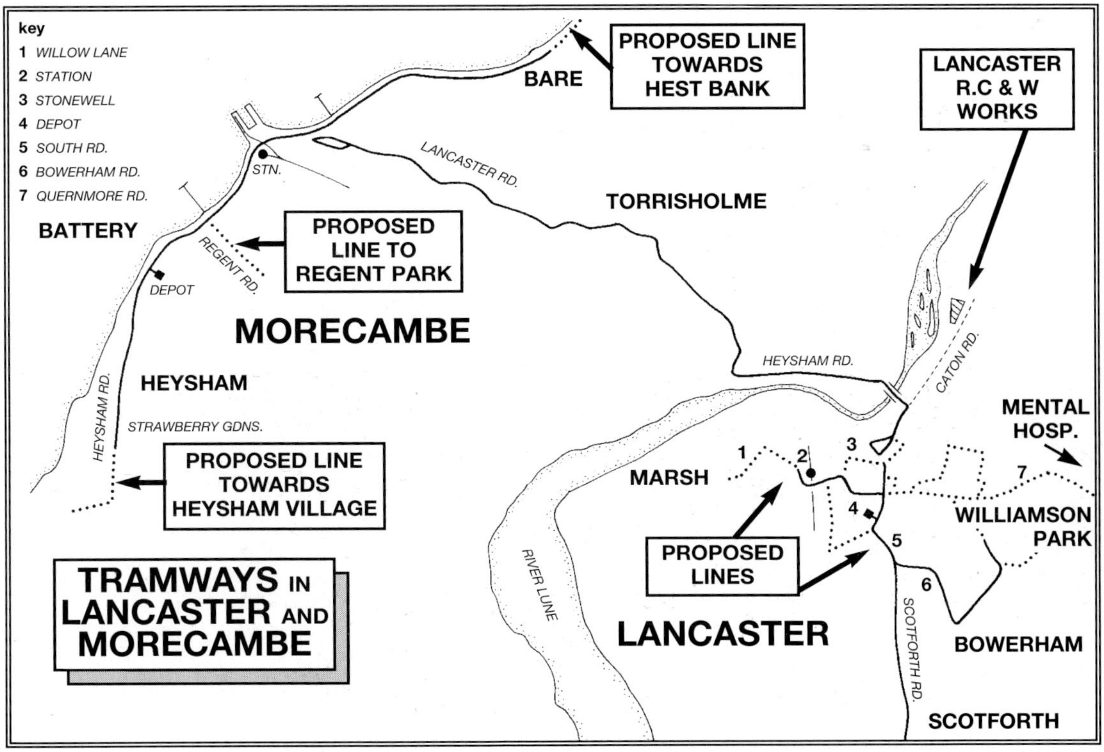

LANCASTER

Already an important railway centre in its own right, Lancaster was jolted into tramway operation in 1899, by the appearance on the horizon of a couple of schemes to bring light railways into the city. Lancaster & District's early (but shortlived) profitable years had reinforced the council view that a system of its own was overdue, and, conveniently, a Bill was to go before Parliament in 1900 to authorise a number of projects in the city. The application for tramway powers was tagged on to it.

Although late promoters of tramway building, the council did not lack enthusiasm, and the planning stages threw up some ambitious ideas, besides attracting outside (and unwelcome) business speculation, whilst sparking a number of disputes with the Lancaster & District Co, which the corporation announced it wished to acquire. This was quickly dropped on realising the precarious state of the Lancaster & District finances. Also dropped from the Bill was an application to build a branch line to the County Asylum (now Lancaster Moor Hospital), which would have been substantial enough to carry goods traffic, especially coal, from the LNWR station. The Lancaster Asylums Board rejected the council's proposed carriage rates, and, it would appear, did not negotiate further. Had this proposed upgrading occurred, another £5,407 would have been added to the £74,057 estimate for the construction of the system.

On 6 August 1900, the Lancaster Corporation Act reached the statute-book. Approval was granted for the construction of 7.69 miles of electrified 4ft 8½in gauge line, over a diversity of routes across the city. This included the connection with the Lancaster & District line at Stonewell. It will be

Below: Lancaster Corporation only ever owned 12 cars, and all except a later pair were purchased for the system's opening in 1902. Built, as expected, by the Lancaster C & W Co, the initial 10 were open-top 41-seaters mounted on Brill 21 trucks and powered by two Westinghouse 25hp motors. Their controllers were also by Westinghouse. Whilst the driver of this unidentified car wears a uniform, his conductor has to make do with his own suit and, of course, ubiquitous cloth cap. The chocolate and primrose livery lasted far beyond tramway days until the formation of the amalgamated municipal transport undertaking in 1974. *Roy Marshall Collection*

recalled that the company had obtained powers to construct a line to the Boot & Shoe Inn at Scotforth, and this was one of the first two lines embarked upon by Lancaster Corporation, the other being to Bowerham and Williamson Park. Both were to terminate in Dalton Square. The two routes diverged at the Pointer on Greaves Road.

A Plymouth man, Mr Tester, was appointed the first Tramways and Electrical engineer. He took up his post on 2 March 1901. At about this time, and before work commenced, a W. J. Kershaw made his début, heading a Birmingham syndicate, intent on acquiring the powers and also the Lancaster & District Co, which it wanted to electrify. It offered the Lancaster & District 18 shillings (90p) a share, but as the corporation would have had to extend the Lancaster & District's Tramways powers for another 25 years to allow the deal to proceed, Kershaw's plans came to naught. The council held the trump card.

Work progressed throughout 1902, and, although everything was more or less ready by the middle of November, it was to be 6 January the following year before the Board of Trade gave the two lines its seal of approval. The Grand Opening came a week later. The lines had cost £25,966 to build, (plus another £1,325 for the depot on Thurnham Street) and they were initially single track, with short double sections in South Road and Bowerham Road being added in 1903 to satisfy the Board of Trade, which was concerned about the 1 in 10 gradient of Bowerham Hill. Ten cars arrived in time for the inauguration of services, all being double-deckers, seating 41, and built, not surprisingly, by the Lancaster Carriage & Wagon Co.

The opening day was quite an occasion for the city, with the mayor, aldermen and councillors proudly riding over the new system and taking turns at the controls, whilst the public flocked to see the new wonders. Services started in earnest the next day, 14 January.

A month or so before, the council had agreed to another line, from Dalton Square up to the LNWR's Castle station and on to Willow Lane, although this latter point, whilst authorised, was never reached, the station remaining the outer terminus. This, looking back, was perhaps inevitable, as the euphoria of January 1903 was to be shortlived. It soon dawned on the astute of the city that its population was insufficient to support an intensive network. In its first year of operation, it made a loss of £2,833, despite a comparatively healthy figure of 1,079,772 passengers. During this first year, the council had made one of its regular bids for the Lancaster & District line and had offered £27,000 in May. Heaven only knows what the deficit would have been had the ailing route to Morecambe come into corporation hands. (The link between the two systems never materialised, either.)

No doubt glad it had not bought Lancaster & District, the corporation could now concentrate on its enlarged system, with the line to the station opening early in 1905, which was also the year when the depot was moved to the other end of the street to allow erection of a new Town Hall to start. With the opening of the new tramway, two more cars were required, but this time the order went to Milnes Voss. The Lancaster Carriage & Wagon Co had been absorbed into the Metropolitan Amalgamated Railway Carriage & Wagon Co shortly after the delivery of the first cars, and its quote for the new pair was reported to be excessive.

Lancaster Corporation Tramways struggled on towards World War 1, unable to turn around losses into operational profits, and when a large munitions works was built in Caton Road, the corporation turned to an unusual, but not unknown, form of vehicle to transport its workers. Five Edison 22-seat, Brush-bodied battery-electric buses were bought in 1916 and 1917. Power for their accumulators was obtained from a specially built generator in Market Square. They were, however, not to last long. After the war, they were tried on a number of routes, but their inability to tackle inclines led to their early withdrawal at the start of the 1920s.

Peacetime conditions gave the tramway no respite. Problems with 'runaways' on Bowerham Hill, and the poor condition of the upper deck of several trams, prompted their conversion to single-deckers, and in an attempt to economise, they were equipped for single-manning. In 1922, the Castle station line was closed and lifted, and two years later the first petrol bus entered service; others soon followed in sufficient numbers to take a share of the Bowerham and Scotforth workings from 1928. This was the year when the council rejected both an approach from Ribble to buy the undertaking and a proposal to convert to trolleybuses.

The cars lingered on, though, for another couple of years, until the Bowerham route closed on 18 January 1930, with the Scotforth service close on its heels on 4 April. Much of what had been hoped for never materialised; the network of lines envisaged in 1902 would never have been sustainable, and so, 39 million passengers, 4,290,000 miles and a mere 27 years later, Lancaster Corporation Tramways, as the last survivor in North Lancashire, closed the tramway chapter of the area's history.

Left: A Lancaster & District Tramway Co ticket. *National Tramway Museum*

Left: No 3 was one of six cars to be later fitted with balcony top covers, increasing the seating by one. This 1927 view at the Scotforth terminus shows motorman Bill Gates at the controls. He was well known in the city. His prominent, yet exposed position at the front of his car was perhaps made more bearable by intakes of the beverage extolled on the billboard to his left! *National Tramway Museum*

Below: 'Coffin Cars' was the irreverent colloquialism for the six remaining cars, which were reduced to single-deckers and adapted for one-man operation between 1917 and 1923. They now seated 24. However, at least the conversion afforded the driver protection from the elements. This could be No 1 of the 1902 delivery. *National Tramway Museum*

CHAPTER 4

THE FYLDE COAST

Forget for a moment the bracing sea air, miles of golden sand, happy childhood memories and all that usually springs to mind about a holiday-centred region. Instead, how about a look at a passenger tramway, which once saw goods trains sharing its tracks; or ambitious plans to cross a wide river estuary by transporter bridge and to lay a meandering line across miles of fields and through picturesque villages; or a fleet of striking streamlined tramcars rubbing shoulders with traditional types? These are all facets of the Fylde's tramways, a good slice of which survives. Appetites hopefully whetted, the picture illustrating the key players can be painted, but first, mention must be made of the pioneer scheme of 1877. A local man, Benjamin Corless Sykes, whose company, Garlick, Park & Sykes, had constructed the Blackpool Promenade from Talbot Square to South Shore, suggested a steam-hauled tramway linking Fleetwood with Lytham. Some 18 miles in length, it would have pre-empted tramcar running along the entire length of coast by nearly 50 years, but was not pursued.

Below: Seen in preservation is Blackpool conduct car No 4 masquerading as sister, No 1. When the system was converted to overhead supply in 1899, all the 16 existing cars were suitably adapted. No 4 became a works car in 1912, hence its longevity. It was built by the Lancaster Carriage & Wagon Co, seating 32, and mounted at first on trunnions, later replaced by ECC trucks. *Author's Collection*

Above: Going north, from a point just below the Cabin, the tramroad entered the first of three urban districts, in which the reserved track was conventional rail on ballast. This is the length alongside Copse Road at Fleetwood, prior to the erection of the 'unclimbable' fence. *Fleetwood Library*

Below: The majority of the Blackpool & Fleetwood Tramroad's track was laid on reserved sections. From the Gynn northwards to the Cliffs the rails were grooved and set into wooden block pavings, as seen here at about the time of opening. This part was, until 1918, the limit of Blackpool Corporation-owned line. The elaborate poles and arms are worthy of note. *Fleetwood Library*

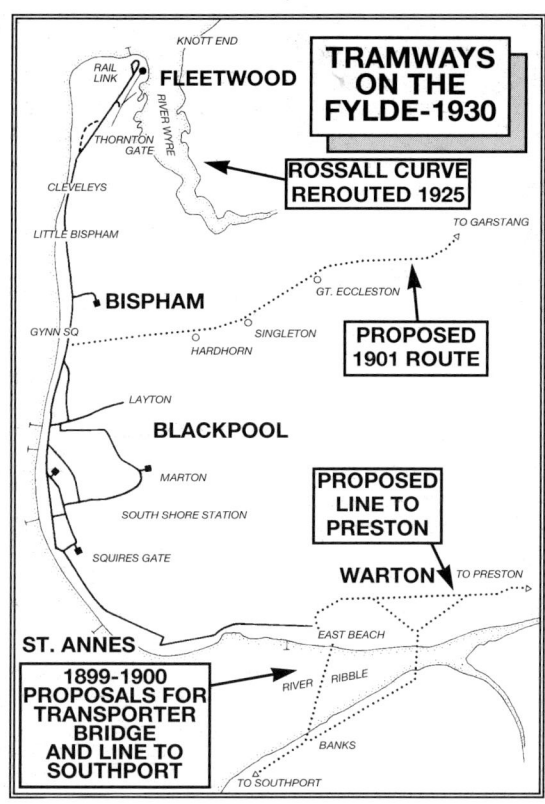

BLACKPOOL

Excited about the prospect of further rapid expansion as the North's premier holiday resort in Victorian times, Blackpool Council was anxious, as the last decades of the century moved on, to provide its thousands of visitors with a means of movement within the town, especially to and from its extremities. It needed, though, to be a mode suited to its image; no dirty steam trams, as Sykes had proposed, nor mundane horse power. Electric traction was the order of the day. The corporation was determined to exploit the vogue, and teamed up with Michael Holroyd Smith, a Halifax tramway developer.

On 3 October 1884, Holroyd Smith demonstrated a standard gauge tram on a test track at his Manchester works to the newly formed Tramway Committee. He had, earlier that year, restored a 220yd electric railway at the Winter Gardens. Although purely a novelty, this little line had impressed Blackpool councillors, and they were once more impressed with Holroyd Smith's discreet conduit system. Discretion was ideal, felt the councillors; no unsightly and potentially dangerous (as they thought) overhead paraphernalia; power collected from within a groove between the lines by a shoe beneath the car was far more acceptable. Tidier it may well have been, but, as will soon be seen, efficient it was not. The Fylde Coast was, and still is, subjected to frequent encroachment by drifting sand and flooding by sea water; far from ideal for the conduit system.

The corporation wasted no time in agreeing to Holroyd Smith's recommendations, and by early 1885, the Blackpool Electric Tramway Co had been formed. There was no shortage of money in the Blackpool of the 1880s and sufficient capital was

Below: Of the 24 Milnes-built tramcars acquired during the Tramroad's first year, 1898, 14 were open cross-bench types, but the remainder were 48-seat 'box' cars, fitted with General Electric Motors and controllers, as exemplified by No 14, seen here in Fleetwood, at a time when the safety of the public was not taken as seriously as it was to be soon afterwards. No protection from the bogies was afforded. The livery was nut brown and cream. *Blackpool Library*

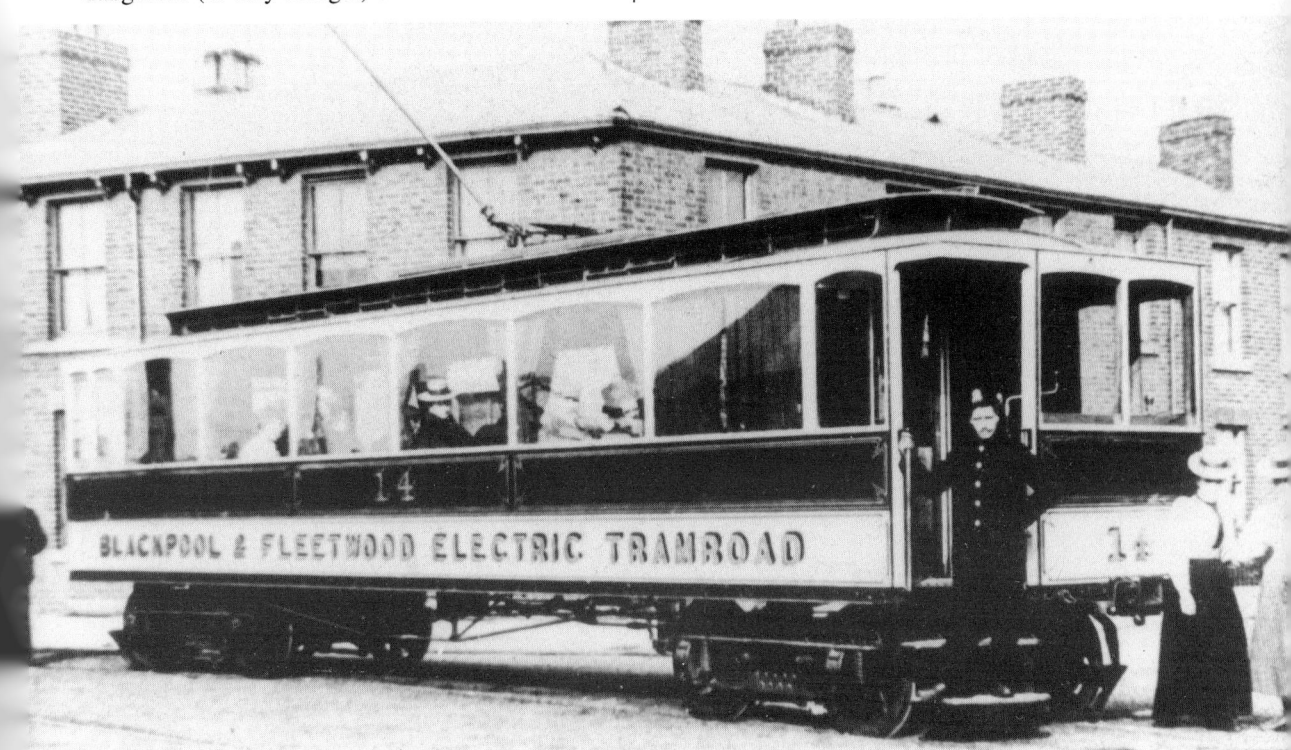

quickly raised to enable work to start on 13 March on what was to be the country's first electric street tramway, along the Promenade, between Cocker Street in the north and Station Road in the south, 1¾ miles long.

Under the Tramways Act provisions, the corporation laid and retained ownership of the permanent way, which it leased to the company, which, on its part, provided rolling stock and built the depot and power station. Holroyd Smith oversaw the work, which included the company's installation of the conduit. All was ready for civic opening on 29 September.

Whilst the corporation had dismissed horse power out of hand in the planning stage, it was these trusty beasts that were to keep the line going in its early days, as the conduit system soon proved temperamental. In fact, the first cars ran over two months before the opening, hauled, with their motors removed or not yet fitted, unofficially, and much to the annoyance of the council, by horses. Ten assorted cars were purchased for the line, and irrespective of demand, that was the number deemed the maximum possible. Apart from the corporation's insistence of double track along its prestigious Central Beach, the rest was single, with passing loops 300yd apart. This, and the heavy voltage loss from the conduit through bad insulation, put a limit on the number of cars able to be out at a given time. A flat fare of 2d was charged, which was a crafty way of minimising use, putting the cars out of reach of the masses.

The cosy relationship between the council and the company soon evaporated, each blaming the other for the delays in getting the system running properly, finding excuses, such as a misunderstanding in 1886 over the use of trailer cars, to drive the wedge in deeper. Nevertheless, the company persevered and some semblance of order prevailed on the line, which was coming under increasing pressure from the resort's growing popularity. The corporation had big things planned for the town: new piers, a big wheel and a tower. The conduit system was in danger of becoming a liability.

Even so, it was extended in 1895 with a double track to South Shore station along Lytham Road and in 1897 from the station back on to the Promenade via Station Road. However, by this time, the company had passed into corporation ownership, the council having taken over in 1892.

The system had been overhauled in 1894, but with a number of council-driven schemes in the pipeline, the most significant for the tramway being the widening of the Promenade, it was now make or break time for the conduit. Seeing overhead

Below: Further cross-bench open cars were added later. This is the last of the 1910 trio of 'Vanguards', Nos 35-7, built by UEC, with a high capacity of 64 and powered by Westinghouse motors (replaced by GEC units in No 37's case in corporation days). Breathtaking rides along the cliff tops were an attraction, but the line could prove too exhilarating at times, hence the roller shutters. Now 'Tramroad' has given way to 'Tramways', although the next delivered, the 1914 'new' box cars, reverted to the former on their side panels. *Fleetwood Library*

Above: Protagonists on the Promenade. The horse bus owners hadn't taken too kindly to the tramway, for obvious reasons, and in pre-corporation days were in almost constant conflict with the company. At Central Beach a conduit car, now corporation-owned and recently converted to overhead power collection, overtakes a horse bus, whilst amply illustrating the massive 20ft trolley poles necessary during the temporary phase prior to the relocating of the tracks on the reserved section. The wire mesh grill to protect upper-deck passengers from the central poles on Lytham Road can be seen. The Clifton Hotel and Roberts' Oyster Rooms are the only surviving buildings on this stretch of promenade, but there is a fair chance that the landau on the left still plies its trade! *Blackpool Library*

Below: A 'Dreadnought' super-tram of the day glides past Central Pier. Two of the subsequent strength of 20-such mammoths were originally company conduit cars. A London man and sometime inventor, George Shrewsbury, patented the 'improved staircase and footboard', which allowed access forward of the bogies, and this idea was adopted for the 'Dreadnoughts' (named after contemporary formidable warships), creating what was at the time a highly unusual front-end treatment. *Blackpool Library*

electrification being adopted elsewhere, the corporation conceded defeat in its battle against the ravages of the sea and decided to convert, Board of Trade permission being granted in June 1898.

All this activity was riddled with power struggles and resignations within the Tramway Committee, no doubt prompted by the frustration of trying to keep the conduit system working. This shambolic state of affairs must have been looked upon with a degree of amusement further up the coast where work was nearing completion on the impressive American-influenced tramroad from Fleetwood down to Talbot Road station in Blackpool. The driving force behind this private venture, the Blackpool & Fleetwood Tramroad Co, was none other than Benjamin Sykes. Assisted by Tom G. Lumb, he used the plans he had drawn up for his 1877 proposal. Knowing they were on to a sure thing, they soon secured financial backing. The northern part of the coast was set to expand rapidly as a residential area, and the railway link between such places as Cleveleys and Rossall and Blackpool was indirect. On top of that, an exhilarating ride along the cliffs would attract visitors, many of whom would travel up to Fleetwood to sample the steamer trips.

The section of the line from the Cliffs to Talbot Road was within the Blackpool boundary, and the corporation were to own this part, leasing it to the company for 21 years. This agreement was reached in May 1896, paving the way for the swift passage of the Blackpool & Fleetwood Tramroad Act, which preceded the Light Railways Act by a few weeks, hence, perhaps, the adoption of 'tramroad' in the title, akin to a light railway. Sykes and Lumb, being developers in their own right, owned much of the land it was to pass over, and the prospectus issued the following June showed the share capital to be £120,000 with authorisation to borrow a further £40,000. It also stated that construction of the line was to be done by Sykes' own firm, the General Tramroad Maintenance & Construction Co, with the contract for laying the 4ft 8½in permanent way and putting up ancillary buildings going to Dick, Kerr and the electrical equipment to be supplied by Mather & Platt of Salford.

Work started at the Fleetwood end on 19 July 1897. As at the Blackpool extremity, the lines were to be laid in the street, but Fleetwood Council, although being involved in the consultative stages, unlike Blackpool, sought no more involvement other than reserving a right to purchase after 30 years. Between Ash Street, Fleetwood, and the Gynn in Blackpool, the tramroad was on reserved track, which included 10 level crossings, and it was protected by an 'unclimbable' iron fence. The main car sheds were at Bispham, also the site for the generating station, the output from which was soon realised to be inadequate for the 8.21 miles of line. Therefore, two accumulator houses, one next to the Fleetwood depot at the Copse and the other at the Gynn, were incorporated into the plans.

On 13 July 1898, the Board of Trade inspected the completed line and permission was granted to open from Fleetwood to the Gynn, but Blackpool Corporation were required to widen Warbreck Road (now Dickson Road) as the cars were five inches wider than in the Order. Operation commenced on the whole length on 29 September, using an initial fleet of 10 cars, which quickly proved inadequate to cater for the crowds who flocked to savour the phenomenon. Under the expert superintendence of the recently appointed general manager, the larger than life John Cameron, the company coped admirably, and an air of professionalism and efficiency soon descended upon the tramroad, a stark contrast to the state of affairs in Blackpool.

However, at least the Tramways Committee there took notice, at first ceasing the practice of employing unreliable casual labour, and by raising wages and recruiting local men. With the news in December 1898 that the council had approved the conversion to overhead current collection, coincidental to a move back into profit, the tramway was poised to turn an important corner, but it was to be far from plain sailing.

To allow work on the Promenade to proceed simultaneously with the conversion, the new poles were erected on the seaward side of the roadway and tracks. They were fitted with ingenious arms which were slotted through holes in them, and off-set, to permit them being slid across when the track was relaid on the new promenade. There had been much local debate about the unsightliness of the poles and arms, and it was decided to keep the arms as short as possible. Enormous 20ft trolley arms were needed to reach the overhead wires, and to make things worse, they had to be off-set on the seaward side of the cars.

On Lytham Road, the anti-overhead lobby, with its opposition to suspended cabling, got its way again, and, despite there being only 4ft 4in between the tracks, central poles were utilised, necessitating the fitting of high wire mesh guards along the sides of the upper decks, to prevent passengers leaning out and banging their heads.

The Board of Trade, one suspects a little grudgingly, approved the conversion as from 21 June 1899, but for a time the conduit cars shared duties with their overhead adapted sisters. Equally important that year was the start made on the

Top: The capability of the 'Dreadnought' to load quickly is shown here, with four, apparently all northbound, attracting plenty of passengers, wonderfully attired in Edwardian period holiday 'togs'. *Blackpool Library*

Above: A line from Talbot Square to Layton was opened in 1902, and No 39, one of the 63-seat double-deckers built by the Midland Railway Carriage & Wagon Co the year before, is seen at the outer terminus in the 1920s. These cars had reversed stairs and were fitted with top covers between 1910 and 1914. *Blackpool Library*

Above: Between 1911 and 1914, 24 toastrack cars were bought by the corporation. Built by UEC, they were fitted with Preston Equal Wheel bogies and had BTH motors and controllers. Although the provision of central gangways in 1936 reduced their seating capacities, they originally carried 69, and were naturally very popular on the Promenade. However, No 70 of the first batch, is seen in Talbot Square about to depart on the Marton Circular Tour. One of the 1902 Hurst Nelson cars, 48, is also leaving the Square for Layton. As it is shown with a short top cover, fitted post-1915, the view can be safely dated as just after World War 1. *St Annes Library*

Left: A notice about fare stages, on the Fleetwood section, issued soon after the take-over of the Tramroad Company by the corporation. *National Tramway Museum*

BLACKPOOL CORPORATION TRAMWAYS.

FARE STAGES
Fleetwood Section.

1d.
Talbot Rd Station & Warley Road
Carshalton Road and Cabin
Cabin and Bispham
Bispham and Norbreck
Norbreck and Little Bispham
Little Bispham and Cleveleys
Cleveleys and Rossall Beach
Thornton Gate and Rossall
Rossall and Fleetwood Road
Fleetwood Road and Stanley Road
Stanley Road and Fleetwood

1½d.
Talbot Rd. Station & Warbreck Hill Rd.

2d.
Talbot Road Station and Cabin
Carshalton Road and Bispham
Cabin and Norbreck
Bispham and Little Bispham
Norbreck and Cleveleys
Little Bispham and Rossall Beach
Cleveleys and Rossall
Rossall Beach and Fleetwood Road
Rossall and Stanley Road
Fleetwood Road and Fleetwood

3d.
Talbot Road Station and Bispham
Warbreck Hill Road & Little Bispham
Bispham and Cleveleys
Norbreck and Rossall Beach
Little Bispham and Rossall
Cleveleys and Fleetwood Road
Rossall and Fleetwood

4d.
Talbot Rd. Station & Little Bispham
Warbreck Hill Road and Cleveleys
Bispham and Rossall
Little Bispham and Fleetwood Road
Cleveleys and Stanley Road
Thornton Gate and Fleetwood

5d.
Talbot Road Station and Cleveleys
Warbreck Hill Rd. and Rossall Beach
Bispham and Fleetwood Road
Little Bispham and Stanley Road
Cleveleys and Fleetwood

6d.
Talbot Rd. Station & Rossall Beach
Warbreck Hill Rd. & Fleetwood Rd.
Bispham and Stanley Road
Norbreck and Fleetwood

7d.
Talbot Rd. Station & Fleetwood Rd.
Warbreck Hill Rd. and Stanley Rd.
Bispham and Fleetwood

8d.
Talbot Rd. Station and Stanley Rd.
Warbreck Hill Road and Fleetwood

9d.
Talbot Rd. Station and Fleetwood

CHILDREN (between 3 and 14 years of age) are Charged as follows:
½d. for 1d. fare
1d. for 1½d. and 2d. fares
1½d. for 3d. fare
2d. for 4d. fare
3d. for 5d. and 6d. fares
4d. for 7d. and 8d. fares
5d. for 9d. fare

Complaints of incivility or inattention on the part of Tramway Employees to be made to

Tramway Offices,
West Caroline Street, Blackpool,
June, 1921.

CHARLES FURNESS,
General Manager.

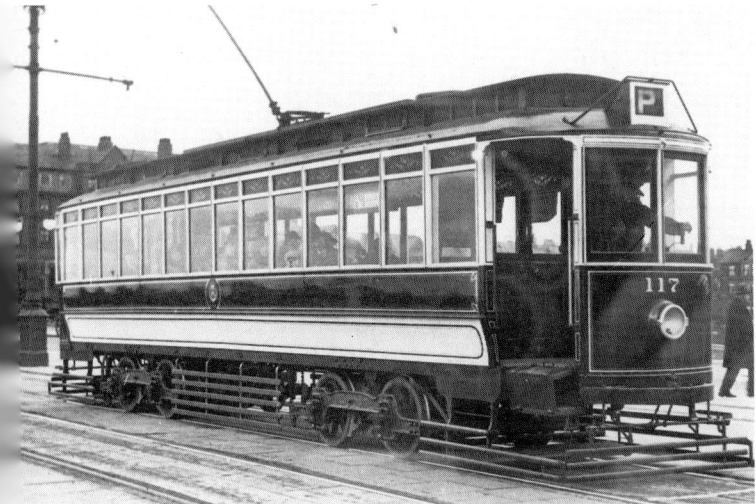

Left: A long tradition in the Blackpool fleet has been the construction of cars for the annual illuminations along the promenade. The second such was the lifeboat, 'Jubilee', converted from Midland open-top car, No 40, in 1926. It had seats for 20 and lasted until 1961. *National Tramway Museum*

Centre left: 1920s consolidation — Blackpool & Fleetwood's former No 29, now corporation No 117, an ER&TCW-built 55-seater combination car of 1899. It had been fully enclosed as a 'glasshouse' car by BCT in the first year of ownership. The 'P' on the roof box denotes Promenade. Others were 'B' — Blackpool, 'C' — Cleveleys, and 'F' — Fleetwood. Lifeguards are now very much in evidence. *Evening Gazette & Herald*

Below: Snapped with a Kodak Brownie 127, Preston Holidays, July 1959, by the author, then 12, is 'Standard' No 160 at Central station. In a six-year period up to 1929, Blackpool Corporation built no fewer than 39 cars. Except for six toastracks, all were double-deckers. This car was completed in 1927, but was not fully enclosed until 1940, as the penultimate in a scheme to provide the Marton route with more hospitable vehicles. As originally turned out, it had open balconies and platforms. Early 'Standards' received the top covers from withdrawn Hurst Nelson cars (Nos 42-53) of 1902, whose numbers they took. *Author*

extension to the Gynn along the new North Promenade. Opening in May 1900, this expansion required more cars, and the two high capacity famous 'Dreadnoughts' of 1898 were supplemented by 10 more. Work was also undertaken at this time on extending the Princess Street depot, southwards to Rigby Road to accommodate 40 cars.

A profit of £7,000 in 1899 was almost doubled to £13,500 in 1900, but such was the demand on the system by the vastly increased patronage, that at times it could not manage, and the West Caroline Street generators on more than one occasion succumbed, requiring current having to be 'jumped' from the Fleetwood company's wires. The corporation's power station was extended in the winter of 1900, not just to cater for these increases, but also for the new line under construction inland to the Marton suburbs, which opened on 23 May 1901.

This new route proved something of a disaster; the rail used had been banned by the Board of Trade and broke the axles on the new cars, which, in turn, were rough riding and totally unsuitable. On top of this, the tracks into the new depot on Whitegate Drive were too tight and had to be relaid. The following year, another new line out to the suburbs, this time to Layton, opened on 19 June, and like the Marton route, never returned a profit; the money was on the Promenade. With great foresight, the corporation recognised the potential, and on 30 May 1902, work got under way on widening it, incorporating a reserved, paved tram track. The final stage was completed in April 1905, eliminating the last single line section and giving private running between South Shore and Talbot Square.

At this time, the Tramroad Co had its eyes on expansion. The hapless Garstang & Knott End Railway had been unable, due to financial problems, to complete its section between Pilling and Knott End. The Blackpool & Fleetwood

Below: From 1927 until 1949, coal trains plied between the railway connection at Fleetwood down the tramroad to the coal merchants' sidings at Thornton Gate. To haul them, an electric locomotive was purchased from English Electric. Powered by a pair of 50hp motors, it lasted until 1963 as a works vehicle, in which capacity it is seen here towing the weedkiller trailer, based on a cross-bench car. *National Tramway Museum*

Right: This railgrinder/snowplough was rebuilt by the corporation from a 1902 Midland Railway Carriage & Wagon Co 'Marton Box' car (number unknown) in 1928. It was still in use in July 1981, when photographed in the depot at Rigby Road. *Author*

Centre right: A bus leaving the road and coming to rest on the reserved track caused this impressive (although not unusual) line-up of patiently waiting cars. From the liveries, especially the green 'V' on the leading English Electric railcoach, the scene is probably from the early 1950s. English Electric provided 45 of these streamliners between 1933 and 1935, with 12 'Sun Saloon' variants in 1939, as part of Walter Luff's five-year modernisation plan. Brush built another 20 similar cars in 1937. *Evening Gazette & Herald*

Below right: To service the suburban Marton route, opened in 1901, a depot was provided on Whitegate Drive. Luff's open-top version of his railcoaches were the 'Boats' and two are seen peering out, alongside No 160, again. The boats' white poles are legacies from the toastracks, used for prominent destination boards, as seen earlier. Marton depot closed along with the route in 1962. A filling station now occupies the site. *Blackpool Library*

Below: A typical Blackpool & Fleetwood ticket. *National Tramway Museum*

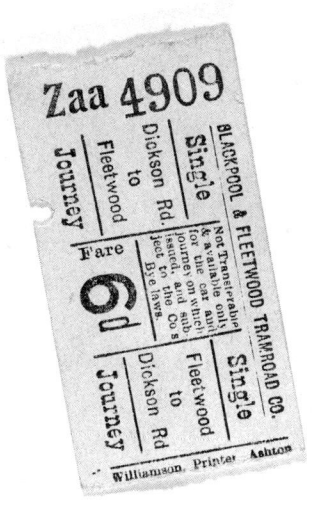

47

Above: The double-deck equivalent of the railcoach materialised in 1934 as this classic style in open and enclosed form. Of the latter, No 259, again built by English Electric, seated 84. Not in original livery, it is, however, in 'as delivered' condition in this early postwar scene. At peak times, two conductors were (and still are) employed; the platform doors are hand-operated by one of them. *National Tramway Museum*

Left: 'Fleetwood Interchange' viewed from the top of the town's famous lighthouse. A Brush railcoach waits for the ferry, on its short trip across the Wyre estuary from Knott End. Had the Blackpool & Fleetwood Co gone ahead with its plans in the early years of the century to acquire the Garstang & Knott End Railway and operate it as a tramway, then no doubt this cross-river link would have seen a much more intensive expansion. Happily, after a couple of years' inactivity, due to problems with the landing stage, the little ferry is still working. *Fleetwood Library*

Right: 'Why should Blackpool have work-a-day trams?' asked Luff. Holidaymakers marvelled at his new cars and enjoyed the comfort of their interiors. This is the inside of an English Electric railcoach. The seats were, of course, reversible. *Blackpool Library*

Below: The Brave New World — once wartime track neglect had been remedied, new cars could be ordered, and Luff again wanted to be a revolutionary, but this time he failed. The 25 'Coronations', built by Roberts of Sheffield, were quite simply, at 8ft wide, too big, and their original Crompton Parkinson 'Vambac' controllers, unreliable. Their 56-seat bodies were mounted on Maley and Taunton HS44 bogies and powered by four x 45hp Crompton Parkinson motors. The controllers were persevered with until 1964, when English Electric units were fitted. However, by 1975 all but one of them had gone. No 660 (No 324 before the 1968 renumbering) is still owned, albeit more or less preserved, though seeing occasional public use. In later years, their chrome work was removed and they received standard livery for the time: cream panels with roof and window surrounds green. They were, nevertheless, handsome machines, and here the first, No 304, has a nocturnal test run to Marton on 8 August 1952. *Evening Gazette & Herald*

Tramroad Co offered to buy the railway, complete and electrify it. It came to nothing, as did another rural tramway idea, that of the Blackpool & Fylde Light Railway Co, which in 1901, as the Blackpool & Garstang Light Railway, had obtained Parliamentary powers to build a line from Gynn Square, out through Hardhorn to Singleton and meander across the Fylde Plain to the LNWR station at Garstang and Catterall. No doubt the almost entirely rural nature of these two schemes saw their downfalls; the subsequent history of the eventually finished but never successful Garstang & Knott End Railway being a case in point.

Blackpool Corporation and the Tramroad Co settled down to moving the wakes weeks masses in their ever increasing numbers, and the similarly expanding local population. Much of that latter increase was brought about by large scale development to the north of Blackpool, and from 1918, the boundary had been extended beyond Bispham, in Tramroad territory. Its lease on the tracks to Talbot Road Station expired in 1919. During World War 1, Blackpool had made it clear it did not intend to renew it, making no secret of its expansionist plans. On 31 December 1919, John Cameron received a cheque for £297,758 from the Blackpool Borough Treasurer in return for seven miles of tramroad, 41 tramcars, three depots and a power station. Ironically, the war had put paid to a plan for the company to enter the town by another route (never determined) and for collaboration with the Lancashire & Yorkshire Railway, probably with a link at Layton (then Bispham) station.

The two systems were gradually consolidated. Early in 1920, the Dickson Road line, which had ended in a single line stub near Queen Street, was joined to the corporation's in Talbot Road. At the same time, the two systems were connected at the Gynn. With the completion of the final phase of the promenade in 1927, between the Pleasure Beach and Starr Gate in the south, through running became possible from there northwards to Fleetwood, 11 miles away.

A most significant and far-reaching decision was made in 1932 when the Tramways and Electricity Departments of the corporation, under the management of Charles Furness, were split, and the post of general manager of the newly formed Transport Department was advertised. The successful candidate was a Yorkshireman, Walter Luff, who took office on 1 January 1933 and set

Below: The 1930s live on — of the 'Balloons', as they became known, 13 were open-top. During the war, however, all were roofed, and No 701, formerly No 238, was one of these. They are easily distinguishable from the rest by their shallower domes. Incredibly, many survive in various levels of rebuilt state; No 701, photographed on 29 July 1981, being in a mid-way condition. Their two-piece destination indicators were early conversions, and rubber glazing has started to creep in. Liveries changed quite rapidly, too, in an effort to disguise the age of these venerable cars. *Author*

Left: Reflecting the then current motorbus trend towards increased efficiency, Brush railcoach No 638 (ex-No 302) was the first Pay-As-You-Enter car, being converted and rebuilt in 1969 by the addition of entrances behind the cabs. These proved too narrow and attracted opposition from the trade unions, and, as a result, No 638 saw little service, being scrapped in 1980 after several years out of use.
Evening Gazette & Herald

Below: No 638's fall from grace was not in vain, though, as from 1972 to 1976 13 English Electric railcoaches were extensively converted into PAYE trams (or OMOs as they were known in the town) to give, in some cases, nearly 20 years' more use. Front entrances were added, this time sufficiently wide, on a frame extension which was to prove troublesome, however, as time progressed. The centre doors were retained as exits, and rubber glazing was utilised. They were numbered 1 to 13 and painted in a distinctive yellow (later cream) and red livery.

Engineer, Stuart Pillar, came to Blackpool from Preston Corporation, where he had been instrumental in that undertaking's famous PD2 to PD3 bus rebuilds, which stood him in good stead for this later project of his. In true Rev W. Awdry fashion, No 6 (formerly No 617) has failed to stop at Permanent Way Works near to the Tower in February 1977 and is well and truly grounded. With the addition of roof box advertisements, there is nothing visible to show it started life as No 270 in July 1935.
Evening Gazette & Herald

Above: With the PAYE principle firmly established, especially for maintaining winter services, attention could be directed towards the ageing 'balloons'. Ideal for moving the summer masses (over 100 passengers with a full standing load and two conductors), their off-season use was uneconomical. As a result, the first 'Jubilee' car, No 761, took to the metals in July 1979. Like the OMO cars it now bore no resemblance to its sisters and its origins as No 725. Its centre doors were replaced by front entrances/exits and its bodywork was completely restyled. The original teak frame was retained, but extended using Duple (Metsec) components. Under the skin, amongst other things, Metalestick suspension (as on the OMOs) was fitted and Westinghouse chopper equipment replaced the series parallel controllers. No 761 is seen on the Promenade on 29 July 1981, by which time work was under way on converting No 714. This time, when this one emerged in April 1983, as No 762, the centre doors had been kept as exits to ease passenger flow. *Author*

about transforming the fleet through a five-year plan, at a time when wholesale abandonment was around him.

'Blackpool is not a work-a-day town, why should it have work-a-day trams?' he asked. Streamlined railcoaches, bedecked in a cream and green livery, replacing the red and white, were his immediate manifestations of change. 'I want to make these cars so attractive that people cannot resist boarding them,' proclaimed Luff, proudly. Double-deck versions (open and closed), 'boats' and single-deck variants followed in number, over 100 in total. Whilst the Promenade service, profitable as it was, seemed ripe for revolution, the loss-making suburban routes needed scrutiny, too. The Layton service was abandoned on 20 October 1936 and replaced by motorbuses (not 'ordinary' buses, though, but rather Luff's streamlined Burlingham-bodied Leyland Titans), and an Ipswich trolleybus was tried on Lytham Road, but nothing came of it. By the end of the 1930s, the 200 or so cars were carrying a staggering 50 million passengers a year, compared with 9 million on the 70 cars of the years before World War 1.

After World War 2, Blackpool shared with its neighbours the inevitable problems caused by the lack of maintenance and the rigours of wartime operation. The Marton route was the worst, and after nearly deciding to convert to trolleybuses, the corporation approved the re-laying of the track instead, and rebuilt cars using Vambac controllers were introduced. In 1949, a rather sad event occurred with the final running of the corporation's coal trains, which from 1927 had been hauled by an electric locomotive from the railway connection at Fleetwood to coal merchants' sidings at Thornton Gate, much to the frequent surprise of unsuspecting sightseers on passing cars.

Track renewal and realignment on the Promenade, to give better clearances, heralded the arrival of new, larger 8ft wide 'Coronation' cars between 1952 and 1954 — Walter Luff's final fling before retirement in the latter year. Sadly, they did not share the success of his prewar designs and proved troublesome. The need for economies, and heavy congestion, saw the end of the Lytham Road route to Squires Gate on 29 October 1961, and the Marton route a year later on 28 October 1962. Twelve months later, again almost to the day, on 27 October 1963, Dickson Road lost its cars. All that remained was the Promenade. At the time of writing its infrastructure is rotting away, victim of the same aggressor that defeated the conduit system. It is to be renewed in a multi-million pound project in partnership with the EU and Lancashire County Council, whilst the tramcar fleet could see rejuvenation with low-cost vehicles, developed by a consortium, which includes Blackpool Transport and East Lancashire Coachbuilders.

Blackpool Council now owns the trackwork, overhead and buildings, the cars being the property of the council-owned Blackpool Transport, which could be sold. The then Transport Minister, Roger Freeman, speaking in October 1993, gave an assurance that any purchaser would be required to continue tramcar operation.

Right: To mark the centenary of the system, eight brand new PAYE cars were introduced between 1984 and 1988. Constructed by East Lancashire Coachbuilders, their bodies incorporated many bus body parts and were mounted on bogies manufactured by Blackpool Transport. No 645 waits at Fleetwood. These cars, along with the two 'Jubilees', maintain the winter service.
Author

Left: Present day on the Fleetwood tramroad: Rossall Beach and a twin-car set with power car No 674 (formerly No 274) leading, heading south. It was rebuilt from a second series railcoach, new in 1934, in 1961, and coupled to a specially-constructed MCW trailer. *Author*

Below: A Blackpool Corporation Tramways ticket. *National Tramway Museum*

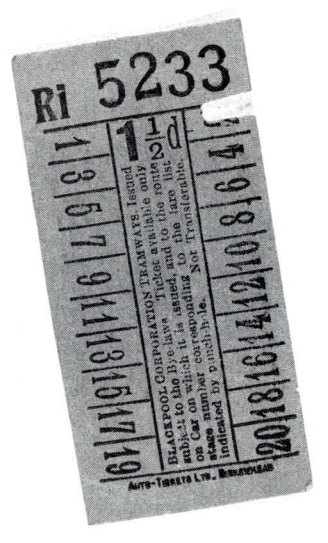

Left: Practicalities, and a need to keep elderly cars running, prompted Blackpool Transport, operators of the tramway since bus deregulation, to make a start on completely rebuilding the double-deckers, whilst retaining their basic profiles. No 723 was the second to be so treated, but differed from the first in lighting detail, skirt trim and by the fitting of rubber bumpers to deter idiots who cling to the back end of moving cars. The 'Tramway' logo is worthy of note, as is the warning by the door, exhorting passengers and pedestrians to take care near the tram tracks. *Author*

LYTHAM St ANNES

What this system lacked in spectacle (its only gradients were to be found at two railway stations and it boasted only five appreciable curves) it certainly made up for in the excellence of its ingenuity, either real, proposed or by association. To the south of Blackpool, what was to become the Borough of Lytham St Annes, grew from 1875 as an extension westwards along the Ribble estuary from the old town of Lytham. Given the title 'St Annes', this new town was a separate Urban District, laid out in garden city fashion in the sand dunes. Mindful of the potential for Lytham and St Annes to join in the Blackpool-centred resort bonanza, powers were obtained in 1880 for a company to operate horse trams from the Blackpool boundary through St Annes to Lytham, but the proposal never left the drawing board.

It will be recalled that Blackpool's South Promenade, enabling coastal access into the town, was not built until 1927, and so when its Lytham Road extension to South Shore railway station was mooted, the possibility of a tramway connection from the south became feasible. As a result, the Blackpool, St Annes & Lytham Tramway Co was formed in 1893 to take over the 1880 powers. By 1896, sufficient capital had been raised and an order made for the construction of its mainly single track, standard gauge line from Station Road to Squires Gate and then through the dunes along Clifton Drive into St Annes. It opened, along with the car sheds at Squires Gate, on 11 July that year, the final section to Lytham being ready for opening in February the following year. A small shed in Henry Street, Lytham, was taken in use at the same time. The company chose not to operate the line directly, but leased it to the British Gas Traction Co Ltd. Although not unique at this time, the system was nevertheless as its operator's title suggests, novel in using compressed coal gas, which fuelled a 15hp Crossley engine, giving the cars a range of about 16 miles.

The gas trams were far from being an unqualified success, and despite a more advanced design of car being bought in 1900, bringing the fleet strength up to 20, their reliability became more and more suspect, to such an extent that for the 1900 season, 20 horse trams were drafted in. It had been the intention to operate on a mixed horse and gas basis (furtive readers' minds could run riot here!), but chaos ensued and delays became commonplace. Furthermore, the company is reported to have become frustrated by Blackpool's refusal to allow it to run over its metals, although at this time there was still no physical connection between the two at South Shore station.

The British Gas Traction Co had already decided to cut its losses and pull out, and in October 1898, a new Blackpool, St Annes & Lytham Tramways Co had been registered, acquiring its namesake's assets for £115,000 along with the Gas Co's rolling stock. Renewal and electrification were, of course, now the priorities and authorisation was forthcoming on 6 August 1900 for the construction of six miles and 35 chains of double and 42 chains of single track. There had been an earlier agreement between Blackpool Corporation and the first company in respect of the leasing of the Lytham Road section, which lay in Blackpool, and the lease was transferred to the new concern, on the proviso that the current for it be purchased from the corporation. The rest of the system was to be powered by electricity supplied by St Annes UDC. (Lytham Council had no public supply.)

Ownership of the proposed line was to be, however, far from settled, as in 1901 the Electric Tramways Construction & Maintenance Co Ltd, a substantial shareholder in its own right, bought the company. In April that year, yet another new company, Blackpool Electric Tramways (South) Ltd, appeared and agreed with the Construction Co, who were to electrify the line, to sell it to them. (It was the South Co who were the early promoters of the line to Preston, referred to in the first chapter, which in May had been refused Parliamentary permission.)

Work on the reconstruction began at last in December 1902, by the Construction Co, with, understandably, the Lytham Road section being undertaken by Blackpool Corporation, although the opportunity was not taken to connect up with its own line at South station. During the construction period, the horse and gas trams had struggled on. The former disappeared, though, in August 1901, and the trailers were sold, the latter's departure from the scene being precipitated by the damage they received when the Squires Gate depot blew down in a gale on 27 January 1903. They were sold to Neath Corporation, who ran them until 1920 on its Skewen-Briton Ferry Line, and to Trafford Park Light Railways in Manchester.

The line effectively closed until 28 May, when it reopened in its electrified state. A clause in the Lytham Road lease prohibited assignment of it to another company, which meant the Construction Co

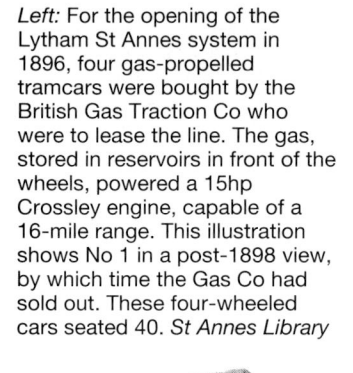

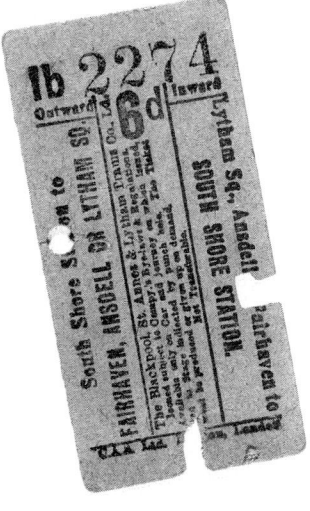

Left: For the opening of the Lytham St Annes system in 1896, four gas-propelled tramcars were bought by the British Gas Traction Co who were to lease the line. The gas, stored in reservoirs in front of the wheels, powered a 15hp Crossley engine, capable of a 16-mile range. This illustration shows No 1 in a post-1898 view, by which time the Gas Co had sold out. These four-wheeled cars seated 40. *St Annes Library*

Above: A Blackpool, St Annes & Lytham Tramways Co ticket, for Lytham Square to South Shore station. *National Tramway Museum*

Centre left: To exploit the Edwardian holiday resort bonanza, cars Nos 21-30 were rebuilt by UEC in 1906 with open cross-bench lower decks, following the success of 10 Brush-built cars of this type delivered the year before. No 23 attracts the St Annes crowds. The car behind is one of the 1-20 series, easily identified by the forward ascending stairs. *St Annes Library*

Left: BEC Car No 12 passes through the sand dunes on Clifton Drive *en route* to Lytham. The drifting sand was later to be a major contributory factor in the decision to abandon the system. Pontin's holiday camp now occupies the area to the right of the view. *St Annes Library*

Right: Twelve larger cars, seating 52, of a supposedly improved design, were acquired by the Blackpool, St Annes & Lytham Tramway Co in 1900. The gas tram concept was never successful and they were all sold in 1903, but, as explained in the text, two other concerns put up with them for some years. An interesting coincidence is the fact that one of them was near to the supplier of the replacement electric cars, as described next. *St Annes Library*

Below: Electrification came in 1903, and what is reckoned to be the first tram into Lytham, which, judging from the posed group, was something of an inauspicious occasion unless of course this bonny bunch had brought the tram down from Squires Gate depot for a ceremony. Car No 5, one of 30 bought in the first year, was built under contract to the British Thompson Houston Co by the British Electric Car Co of Trafford Park, Manchester and mounted on that manufacturer's SB60 truck, with two GEC 25hp motors and BTH B18 controllers. *St Annes Library*

57

had to discharge the Blackpool, St Annes & Lytham Tramway Co's liabilities, paving the way for the latter to operate the system, and for the other two to be wound up.

So much, for the time being, about the reality. Apart, of course, from the exception of the gas trams, the Lytham St Annes story so far has not been particularly amazing, but with one idea it was to be associated with, it came close to making 'spectacular' look like the all-time understatement of the tramway age. '... the maddest tramway never to be built' was how A. Winston Bond described the plans when writing in *Modern Tramway* in 1969. As already mentioned, there were proposals to extend eastwards from Lytham through Freckleton to Preston, and in September 1903 a 1,100yd single line was built from the Lytham terminus to East Beach, as the first part of the scheme.

However, there was to be more to it than just a rural ride to Preston, for at Warton, another line was to diverge southeasterly across the marshes and then due south to Hesketh Bank and southwestwards to Southport, necessitating the crossing of what, to this day, creates a 30 miles round trip, the River Ribble.

Disregarding several road and present day motorway proposals, that of George Binnie, the eminent Glasgow engineer, for a 'rail-plane' bridge crossing in 1927 and the shortlived idea in 1896 by a Southport architect, S. Speedy, for a swingbridge, there were actually two tramway projects. The first was based on Volk's Brighton-Rottingdean concept, with tramcars driving on to a platform with long legs, which ran on rails on the riverbed. On reaching the navigable channel, the cars would then run on to a transporter bridge. Unfortunately, the Southport & Lytham Tramroad Bill, presented to Parliament in 1899, failed in the Lords — Preston Corporation, with its newly opened docks, successfully scuppering the plans,

Below: English Electric supplied Lytham St Annes Corporation (as the undertaking was by then known) with 10 balcony cars in 1924. Peckham trucks were specified, the corporation still happy with four wheels for its cars. They seated 61, and remained in this open condition, as depicted by No 43, until withdrawn on the closure of the system. *Evening Gazette & Herald*

Top: With the opening of Blackpool's South Promenade, enabling the 'Blue Cars' to work between Squires Gate and the Pleasure Beach, six second-hand cars were acquired in 1933 and 1934, introducing fully-enclosed trams to the fleet, along with a jazzed-up livery, no doubt in response to the arrival of Blackpool's streamliners. First arrivals were four single-deckers (another innovation), from the Dearne District Light Railway, built by English Electric in 1924. This is No 54. *National Tramway Museum*

Above: Next was No 55, formerly Accrington No 39, a Brush product, already seen in Chapter One. This was destined to be Lytham St Annes' only car mounted on bogies.
National Tramway Museum

on the grounds that the bridge would be a danger to shipping, and the whole contraption would silt up the channel.

The promoter of the scheme, The Southport & Lytham Tramroad Co, bided its time for 12 months and tried again in 1900, with revised plans for just a transporter bridge, some two miles upstream, where the river is much narrower than at the previous point. This time, the bridge would be reached by the branch off the Preston line referred to, and on leaving it, cars would cross an iron pier, due south to Hesketh Bank, and then run southwesterly to Southport. The transporter bridge itself was to be 60ft long and would take one minute to cross the channel. The total cost was estimated at £183,500.

The Bill became law, but all that was ever completed was the East Beach extension. Financial backing was not forthcoming and the powers lapsed in 1910, sadly, perhaps, as the second proposal was certainly a practical one, especially in view of the success of the then recently opened transporter bridge between Widnes and Runcorn.

A useful precedent for the Blackpool, St Annes & Lytham Co was set in 1904, when Parliament granted compulsory running powers for a company to operate over Newcastle-upon-Tyne Corporation's tracks, and the company took this as an opportunity to resolve its problem of Blackpool's refusal to allow through running north of Station Road. Being well aware of the cost of promoting a Bill, it was decided to apply pressure on its neighbour, with the ultimate threat of taking its grievance to Parliament. Three alternatives were put to Blackpool: (a) To allow through running along Station Road to what is now South Pier (then Victoria), for a payment of £250 per year, with the company to run all its cars without any charge being levied for current; (b) to run along the top end of Lytham Road to the Manchester Hotel, with the corporation taking one penny for each through passenger and the corporation paying the company 13s 4d for each car operated north of South Shore station; and (c) to run along Lytham Road to the 'Royal Oak' and inland to Central Drive and up to a point north of Central station, with similar financial arrangements to (b).

To the company's delight, Blackpool agreed to all three, and through running started in July 1905, on the joining of the tracks at South Shore station. However, the corporation were informed that running powers along the Promenade to Gynn Square would still be pressed for. (There was a reciprocal arrangement for corporation cars to run to Lytham, but this was rarely, if at all, ever exercised.) It will be recalled that the company leased only the lower Lytham Road portion, and in 1917 this expired, with Blackpool taking over the section as far as Squires Gate, but allowing the company to retain running powers.

Below: No 23 again, passing St Annes Library on its way to Lytham. The paving blocks between the 92lb per yard rails can be clearly seen. Some of these were balastic lava, whilst others were either Jarrah or white wood. These are probably the latter. The gentlemen's fashions date the photograph around the time of the St Annes council take-over of the company in 1920.
St Annes Library

On 28 October 1920, amid much civic celebration, the company passed into council ownership, St Annes UDC having agreed to pay £132,279 for it, after 12 months of negotiation and eventual Parliamentary approval. Lytham UDC had initially been party to the process, intending to purchase its part and then lease it to St Annes, but decided to let St Annes get on with the acquisition alone. It was to be only two years, though, before Lytham entered the frame again, but this time as part of the new borough on the amalgamation of the UDCs in 1922. In the November of that year, the new Tramway Committee had its inaugural meeting, when it was informed that through running along the Promenade to Adelaide Place, near to North Pier, had been acceded to by Blackpool. Lytham St Annes then abandoned through running along Central Drive, all its cars operating the full length of Lytham Road and then up the Promenade, from 22 January 1923.

This was also the year when the inevitable motorbuses crept on to the streets of the borough, four Guy single-deckers being introduced on a service between St Annes and Lytham, avoiding the tram route. (A charabanc had been owned briefly in 1913.) Other routes followed, and by February 1928, 13 buses were in stock. However, there was no question of abandonment. Indeed, the system was reported to be in good shape; both the track and the cars, which now had reached the goal of Gynn Square (as from 19 July 1926) and were also to be seen occasionally as far north as Bispham, on hire to Blackpool. The ultimate was attained, though, in the summer of 1928, when Lytham St Annes 'Blue Cars' as they were known made their first journeys over Blackpool's New South Promenade, between Squires Gate and the Pleasure Beach, and then up to the Gynn. In the October, powers were obtained finally to abandon the Lytham Beach extension, which had been motorbus-worked since 23 January 1926. The line was then lifted.

Despite this reduction in route mileage, traffic increased, mainly due to the new services along the South Promenade, and, as a result, six second-hand cars entered the fleet in 1933 and 1934 — four single-deckers from Dearne District Light Railways, and double-deckers from Accrington and Preston.

'ABOLISH THE TRAM — Thousands of pounds could be saved in operating costs — COUNCILLOR HORSFALL'S STATEMENT' was the banner headline in the *Lytham St Annes Express* on 9 April 1936. The councillor put forward a convincing argument for the diesel bus, quoting the wise move of the Joint Traffic Board of Colne, Nelson and Burnley (sic) in abandoning its trams. The lights were already going out on the system; Blackpool had made a bid for it the year before and the problem of drifting sand along Clifton Drive was getting worse. Then another proposal, for Blackpool to acquire the track as far as St Annes and extend its reserved section to there, also came to nought. On 15 December 1936, the Lytham-St Annes section closed, with St Annes-Squires Gate converted on 28 April 1937 to 'Gearless' Leyland Titans.

Blackpool trams continued to rub shoulders on the Promenade with Squires Gate-shedded vehicles after abandonment. These were still blue and after 1974 were owned by Fylde Borough Transport. With more than a touch of irony, these successors to Lytham St Annes regularly worked through Freckleton to Preston; *déjà vu*!

However, history was to repeat itself yet again, as in June 1994, Blackpool Transport successfully bid for 'Blue Buses', with the intention of operating the company as a subsidiary. It had, at last, got its way.

Below: Joint working from the Gynn to Lytham shown on this ticket issued by both operators. *National Tramway Museum*

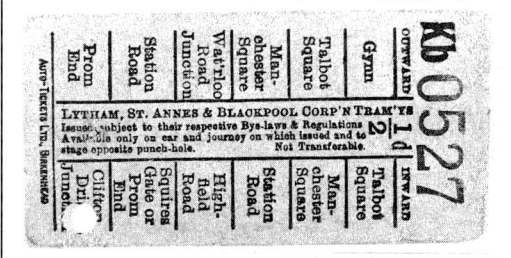

Below: Completing the secondhand cars was Preston's No 42, one of three built by that operator, as will be described in the next chapter. It was given the fleet number 56, the last for Lytham St Annes. *National Tramway Museum*

Above: Sadly, this old postcard view of Garstang Road, Preston, near to what is thought to be St Thomas's Road, cannot be accurately dated. However, it is more than likely to be post-1886, when the Preston Tramways Co sold its route to Fulwood to the corporation. This had been the pioneer line in the town, laid in 1879 to 3ft 6in gauge, and horse-worked. *Author's Collection*

Below: In 1882 Preston Corporation constructed two more routes, to Ashton and from Penwortham, through the town centre and then along New Hall Lane to Farringdon Park. The operation was leased to Harding's, a horse bus operator. Here a car is at Fulwood Barracks, the terminus of the old company line. The 'knife-board' seating, popular at the time, can be clearly seen on the open top deck. *Roy Marshall Collection*

CHAPTER 5

PRESTON & DICK, KERR

Once every Preston Guild' is a well-worn expression in Lancashire, which originates from the extravaganza held in the town every 20 years. More correctly the 'Guild Merchant', it can trace its roots back to medieval times and was of religious origin, but later became associated with craftsmen and traders. In more recent times the occasion has been used as a 'shop window' for industry and commerce, while retaining a strong religious element. Of course, the whole week has become something of a festival for local folk and attracts not only exiled Prestonians back to home, but visitors from all over the world, and in 1882 420,000 arrived by train. Those eager day-trippers would have left the station and, on stepping into Fishergate, been able to gaze upon what had prudently been timed to coincide with the Guild, the opening of the town's new horse tramway.

Constructed and owned by Preston Corporation, the new system augmented an earlier, much smaller one, opened in March 1879, by the Preston Tramways Co, the 1870 Tramways Act precluding the corporation from active involvement in its operation. The concept of a horse-drawn tramway was, nevertheless, not new to the area, as in 1803 an intriguing line was built to connect the two separate sections of the Lancaster Canal, divided by the valley of the River Ribble.

Carrying freight only, it closed in 1864 and much of it has now disappeared, but a fair length, dead-straight southwards from the river, is now in everyday use as a public footpath. However, the Tram Bridge over the river, at the foot of what was the Avenham incline, is, in fact, a replacement replica of the original all-timber structure. Also surviving, happily, as it had been intended to fill it in during the Fishergate Shopping Centre development in the 1970s, is the tunnel on the approach to the Preston basin, now an access to the car park.

Of some two miles in length, the pioneer 3ft 6in gauge passenger-carrying street tramway's route started in Lancaster Road, in the town centre, and traversed North Road and Garstang Road (now the A6), northwards to Victoria Road just outside the town's boundary, in what was to become Fulwood Urban District, and then eastwards to the Barracks. It was adjacent to the terminus, near where the Garrison Hotel stands, that the depot was sited.

The service, in part, replaced an earlier horse bus operation, started initially in 1859 by a Richard Veevers, who soon sold out. About 12 months later, another sale saw a new company formed, which expanded operations by introducing a new route to Broughton, some three miles further north from Fulwood. In 1870, or thereabouts, the company was taken over by another horse bus proprietor, Harding, who had services to Ashton on the Blackpool Road, and, it is thought, to Walmer Bridge and Much Hoole in the Southport direction and to Higher Walton on the 'old road' to Blackburn. Harding was not to rest with just his omnibuses, though, and was to become involved with tramways, as it was he who leased the new system from the council, under the Preston Corporation Tramways name in 1882.

It involved two routes. The first, three miles long, ran from the Pleasure Gardens in New Hall Lane (near to the present-day Farringdon Park bus terminus), into town by way of Stanley Street and along Church Street. The route then continued along Fishergate, passing the stables and car shed, on what is now the County Hall site, down Fishergate Hill and terminating near to where the new Penwortham Bridge was to be built in 1912.

The second served the expanding township of Ashton-on-Ribble. It left the Town Hall by Friargate, and reached its outer terminus on Blackpool Road (then called Long Lane) at the Newton Road junction via Tulketh Road, Wellington Road and Tulketh Avenue, a journey of two miles. For both services, eight double-deck cars were needed. The gauge was also 3ft 6in.

In 1886, the Tramway Company sold its Fulwood operation to the corporation, Harding now working the whole system of three routes, and in 1895, following acquisition of the land to build

County Hall, a new depot was built in Old Vicarage, off Lancaster Road. The horse trams survived the turn of the century, Harding's lease expiring on the last day of 1903.

The electric era loomed closer for Preston's trams with preparations for the corporation to take up its Tramways Act option and the take-over of the system completely. Two Acts of Parliament, in 1900 and 1902, paved the way for another new but considerably expanded system, to be powered by electricity. Under the first piece of legislation, 17 tramways were authorised, although six were, in reality, short sections facilitating crossovers between the routes in the town centre, or part routes. A gauge of 4ft was agreed, as was the insertion of a clause allowing the doubling of single line sections later, and facilities for joining up with any other systems entering the town. Tram sheds were to be provided near to the Dick, Kerr tramcar works in Strand Road. All in all, 11 miles, one furlong, 8.59 chains of track was to be laid (excluding crossovers), with six miles, five furlongs, 2.65 chains of it being doubled.

Despite Harding's lease expiry date getting closer, no work was started, and in 1902 the second act superseded the first. Now 24 tramways were authorised, many of them, of course, as in the 1900 Act, being part routes or just crossovers, but the main difference was the inclusion, omitted previously, of the Fulwood lines. All were now to be laid to the standard 4ft 8½in gauge, giving a total route mileage of 14 miles, five furlongs, 2.5 chains. Over half, (eight miles, four furlongs, 9.4 chains) was to be double track.

Being mindful of the cessation date for Harding's operations, work did not begin until early in 1903 and the depot and generating station in Holmrook Road, Deepdale, were finished in March 1904. (The 1900 site proposal had been dropped.) Accommodation for 30 cars was provided on six internal roads, whilst the power station contained two x 300kW Dick, Kerr 500/550V generators, with engines by the Bradford firm of Cale Marchant & Morley, fired by two traditional Lancashire boilers. Ancillary equipment included a storage battery for use in the event of a breakdown, as well as the mechanism to feed in extra steam from the adjacent Corporation destructor plant, a facility of which good use was made from time to time.

The first 30 cars had been ordered, naturally, from Dick, Kerr in October 1903. At £472 each, the cars were all double-deck, without canopies or vestibules.

So the cars arrived, but the track and overhead were far from ready, delays having been caused by the failure of the council to agree on the tender from Dick, Kerr. This may sound surprising, but perhaps even more so was the reason: Dick, Kerr

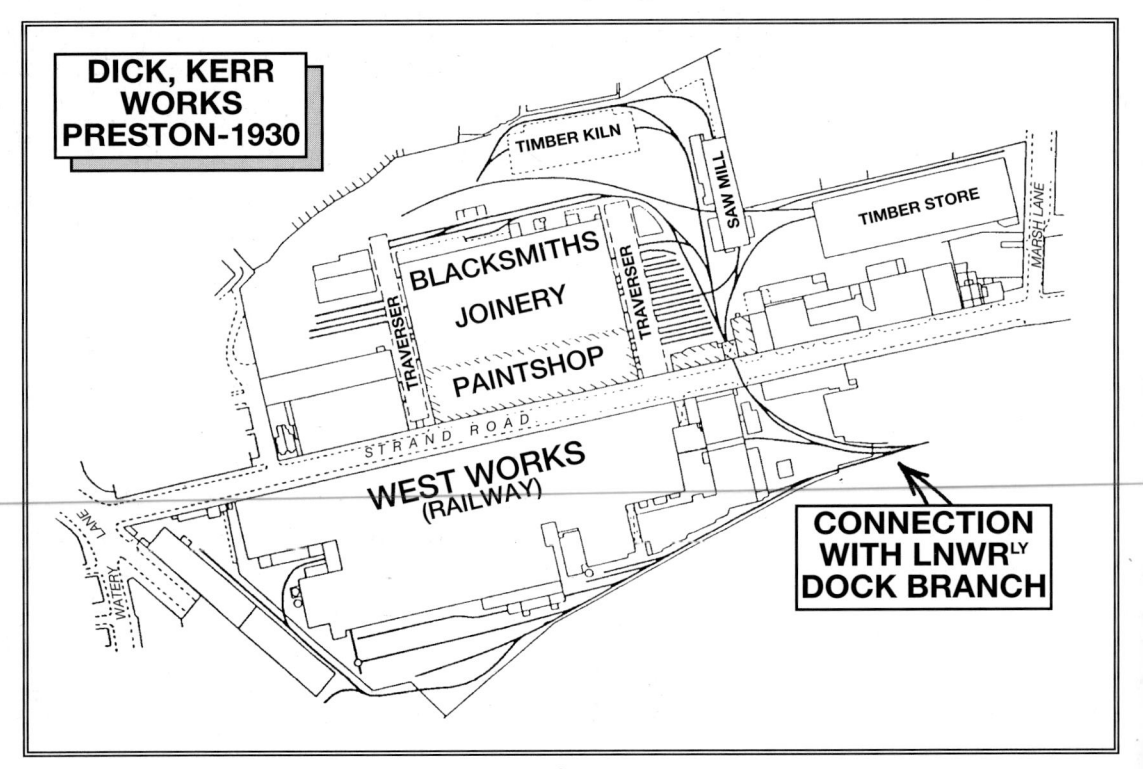

Above: This chapter concerns itself not just with Preston Corporation, but also with Dick, Kerr and its former and subsequent associated companies, based in the town. Two years before the Preston electrification, Dick, Kerr, incorporating the Electric Railway & Tramway Carriage Works and English Electric, exhibited at the 2nd International Tramways and Light Railways Exhibition at the Royal Agricultural Hall in Islington, London in July 1902. Apart from all the aspects of complete tramway construction, for which the concern was renowned, a cinematograph of the Preston Works was shown. *National Tramway Museum*

Right: The tram visible in the middle distance was an ex-LCC conduit car and this is the interior of it, to standard 'Preston' design, adopted by Preston and many others. *National Tramway Museum*

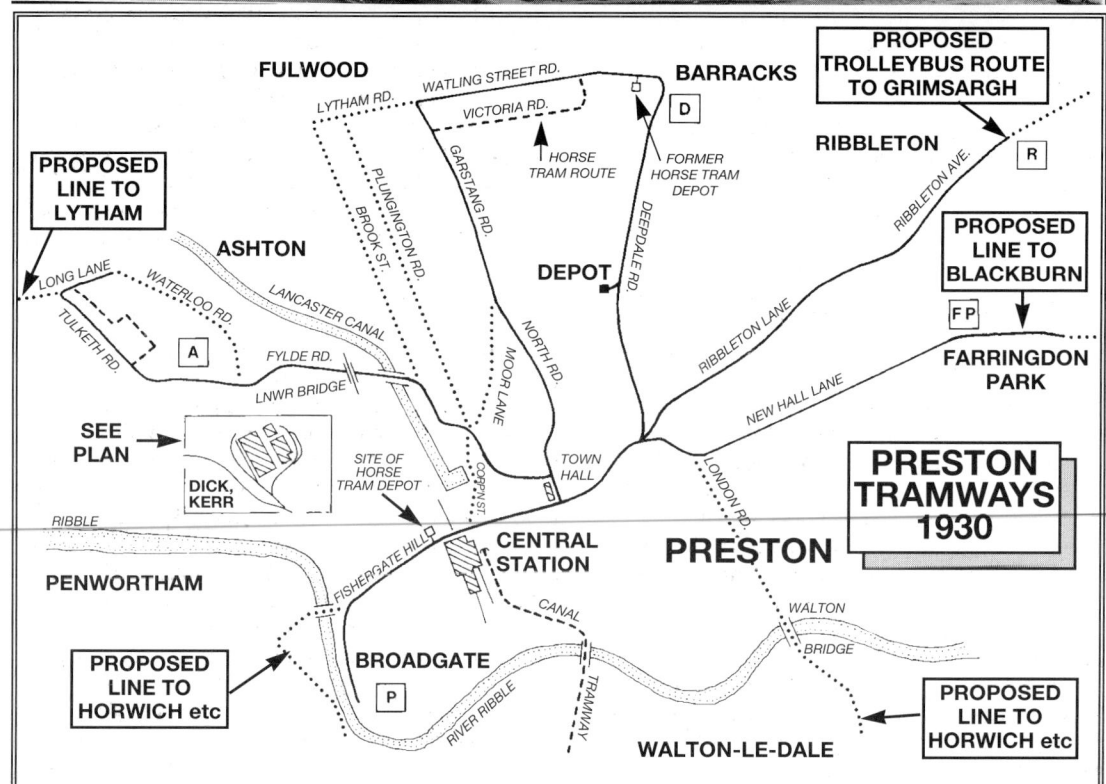

PRESTON CORPORATION TRAMWAYS

Map: Preston Tramways 1930

Key locations and features shown on map:

- **FULWOOD** — Lytham Rd., Watling Street Rd., Victoria Rd.
- **BARRACKS** — Horse Tram Route, Former Horse Tram Depot [D]
- **PROPOSED TROLLEYBUS ROUTE TO GRIMSARGH** [R]
- **RIBBLETON** — Ribbleton Ave.
- **PROPOSED LINE TO LYTHAM** — Long Lane
- **ASHTON** — Waterloo Rd., Tulketh Rd., Plungington Rd., Brook St., Lancaster Canal, Garstang Rd., Moor Lane, North Rd., Deepdale Rd. [A]
- **DEPOT**
- **PROPOSED LINE TO BLACKBURN** [FP]
- **FARRINGDON PARK** — Ribbleton Lane, New Hall Lane
- **SEE PLAN** — DICK, KERR
- Fylde Rd., LNWR Bridge, Site of Horse Tram Depot, Copp St., Town Hall, Fishergate Hill
- **RIBBLE**
- **PENWORTHAM**
- **CENTRAL STATION**
- **PRESTON**
- **PRESTON TRAMWAYS 1930**
- **PROPOSED LINE TO HORWICH etc** — Broadgate [P]
- **WALTON BRIDGE** — London Rd., Canal, Tramway, River Ribble
- **WALTON-LE-DALE**
- **PROPOSED LINE TO HORWICH etc**

Left: With the electrification and subsequent enlarging of the system, by Dick, Kerr, naturally, Preston Corporation took delivery of an initial batch of 30 cars in 1904, all built by the ER&TCW Co in the town. This is thought to be No 1, the first of 26 four-wheel cars, seating 48, recorded by the works photographer. At this time, a fully open construction was becoming a little dated. *Roy Marshall Collection*

had sub-contracted a small part of the works to an American firm, Lorain, which obviously did not suit. The matter was resolved, eventually, with Dick, Kerr getting the contract for the overhead and most of the trackwork (using BSS 2 and 7 rail), but with some specialist work on points etc, going to Hadfields.

Depending on how interpreted, there were to be either six routes or five, as those to Fulwood and to Deepdale were, in 1905, joined at the Barracks and worked as an inner and outer circle. Fulwood was now growing rapidly as a suburb, so Watling Street Road in its entire length was used for the new electrified route, as opposed to that of the horse trams, along the parallel, but narrow, Victoria Road.

The horse trams to Penwortham (a rather deceptive destination, as that township is actually over the Ribble in what is now the Borough of South Ribble) were to be replaced by the electric cars on a route extended along the riverbank (Broadgate), and the route along New Hall Lane was to be directly replaced, although to the by now named area of Farringdon Park. The longest, to Ashton, straightened out the horse tram route, which tended to wander through the district, and terminated in Waterloo Road. (There had been an intention to build this one as a circular around the suburb, to be worked in both directions, but the idea was not carried out.) The final proposed route was to be to Ribbleton, but more about that in a moment.

The inaugural day of the first two routes, to the Barracks at Fulwood, via North Road and Watling Street Road, and to Farringdon Park was 7 June 1904. Of course, Harding had worked his last horse tram on the previous New Year's Eve, which necessitated a temporary horse bus service over the former tram routes, operated by Harding, who now saw himself, albeit briefly, as a horse bus operator again in the town. (Horse buses had, however, continued out of town.)

Next to open was the Deepdale route, on 30 June, although initially the Barracks were not reached, the cars reversing at Moor Park Avenue. On the same day, the first cars ran to Penwortham (Broadgate), and the Ashton service commenced on 9 July.

What of the Ribbleton route? This had rather a painful birth. At the March 1904 council meeting, a decision to defer its construction for 12 months to save money was challenged, and public opinion, bordering on outrage, was such that the Tramways Committee's recommendation was overturned and work was given the go-ahead. Once the rest of the routes were opened, work started on laying tracks and erecting overhead along Ribbleton Lane to the Bowling Green Hotel on the road out to Longridge. The first cars ran on 26 January 1905. In April that year consideration was also given to starting the approved London Road route to Walton. No decision was made, but the matter would not rest, as Walton-le-Dale UDC over the ensuing months (and, as will be seen, years) maintained pressure on Preston Council. Their efforts, though, were to come to naught.

An interesting aspect of working occurred in the Town Centre, where, in Ormskirk Road and Church Street, 'blind' sections which created bottlenecks were controlled by automatic signals.

It was not until the completion of the introduction of single-crewed buses as late as 1986, that a tradition of service linking through the Town Centre came to an end, and it is probable that the practice started from the outset of the electric car operation. Ribbleton cars ran through to Ashton, and Penwortham to Farringdon Park, (carried on by the buses until a change of linking in 1953) or to Withy Trees, on the outer circle route at Fulwood. In respect of the circles, the inner left town by way of Deepdale Road, along Watling Street Road, with the outer, the other way round, leaving town along North Road and into Garstang Road, turning right at the Withy Trees into Watling Street Road. Some services to Ashton started from the railway station, as did occasional short-workings up Fishergate into the town centre. Other short-workings from the station and town centre were football specials to Deepdale, and to such points on the circles as Moor Park, Sharoe Green, the Barracks or Withy Trees, to Powis Road at Ashton, Skeffington Road and the Cemetery on the Farringdon Park route and to the 'Old England' on Ribbleton Lane. Mnemonics were a Preston tradition for route identification: Ashton — A; Inner Circle — D; Outer Circle — F; Farringdon Park — FP; Penwortham — P; Ribbleton — R; Other Workings — O.

As the system settled down, minor adjustments were made as necessary; in 1907 Ribbleton terminus benefited from the installation of a trolley reverser (later experiments with bow collectors were not pursued), all cars were fitted with Simpson

Above: No 6 of the same batch pictured on the Fulwood circular route. As has been observed before, the conductors (one perhaps a trainee) did not have the luxury of uniforms. *Author's Collection*

Left: The railway bridge in Fylde Road on the Ashton route proved troublesome, well into the motorbus era, although the first, open-top cars negotiated it without trouble. However, precautions had to be taken. The rule book, later issued to staff, stated 'Before passing under the LNWR railway bridge in Fylde Road, cars must be brought to a dead stop. On open cars the conductor must warn all top deck passengers to "Keep their seats and avoid touching the trolley wire" and must then signal the driver "Right Ahead" and remain on the top deck of the car during its passing under the bridge.' On a test run, No 16 is watched by officials and a Preston Borough Police Sergeant. *Author's Collection*

Right: A view of the depot at Deepdale in the first year of operation. Six tracks were provided, all with pits, the workshops and power generating plant being situated to the left-hand rear of the picture. *Author's Collection*

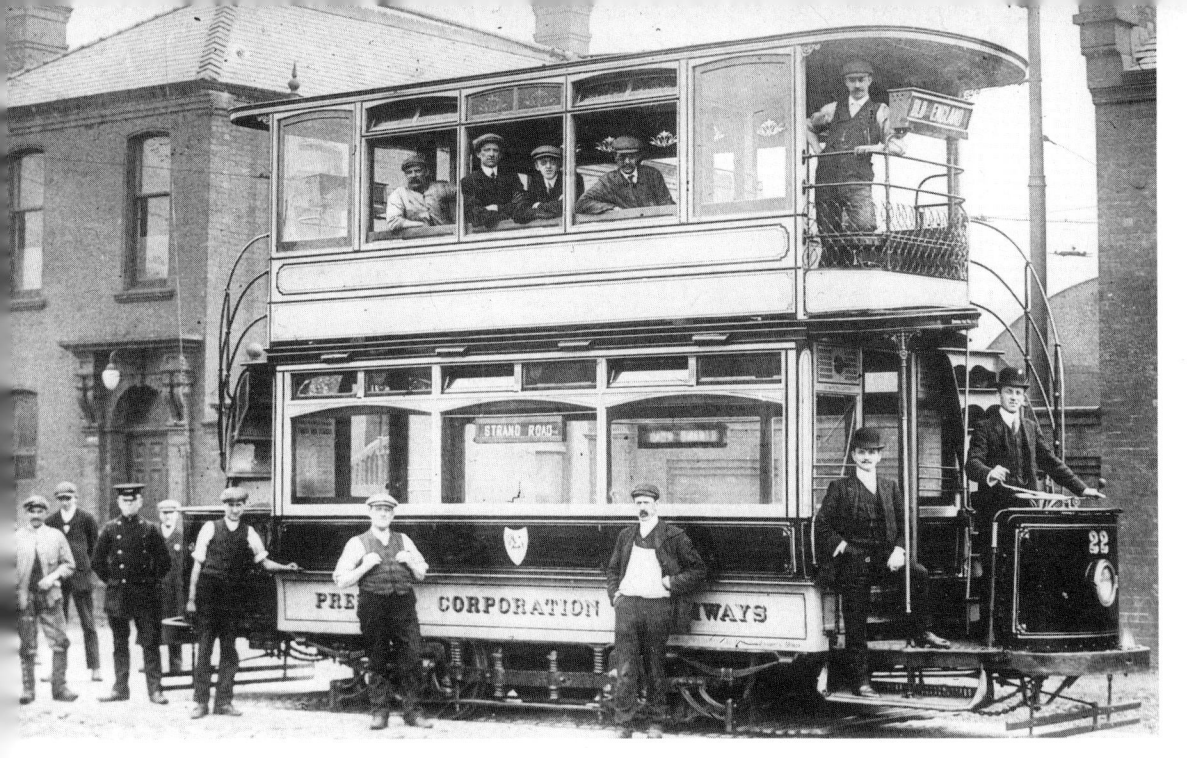

Above: Eighteen of the first four-wheel cars eventually received UEC short top covers between 1907 and 1913. Here is No 22 at Deepdale, with the original offices behind. These were demolished in 1914 to make way for the first extension and the provision of new offices on Deepdale Road, still in use at the time of writing. 'Old England' is displayed on the blind, this being a short-working on the Ribbleton route. *National Tramway Museum*

truck brakes, invented and developed by the manager, J. F. Simpson, starting in 1908, and in 1909 the opportunity was taken to tidy up the junctions at Long Lane/Tulketh Road and Garstang Road/Moor Lane, the unused points being lifted. Needless to say, the Walton issue raised its head again and, once more, to no avail!

Early in 1912, three new single-deck cars were ordered and delivered later in the year from Dick, Kerr, and a tender was accepted for a temporary shed extension. By September that year, a rolling programme of fitting top-covers to the originally open-car fleet, reached 20. It was deemed unwise to treat the remainder because of the infamous problem on the Ashton route — Fylde Road railway bridge with its restricted clearance. Not until 1957, when the roadway under it was lowered, did the problem go away, and obviated the need for special provisions. In later motorbus days, 'lowbridge' double-deckers, as explained in due course, had to be used and there had even been an agreement to purchase 'Beverley Bar' type vehicles.

Apart from the renewing of the heavily used Church Street trackwork, 1913 also saw a reorganisation of services, prompted by a wish to operate covered cars as much as possible. Therefore, the Ashton route, on which a Pay-As-You-Enter experiment was tried at that time, became mainly single-deck worked, its former linking with Penwortham going to the Ribbleton service. Penwortham-Withy Trees was discontinued, this destination being served as a short-working on the Outer Circle, when needed.

Later in the year, the corporation decided upon three new routes and agreed to seek Parliamentary powers. The proposed routes were, firstly, from Garstang Road along Addison Street to Ashton; secondly, from Corporation Street to Withy Trees via Fylde Road, Brook Street and Lytham Road; and finally, what had become the old chestnut, from Stanley Street to Walton along London Road. However, by the time the legislation reached the statute-book, the Brook Street line had been moved to the parallel, but far narrower and really less suitable, Plungington Road, and the Ashton via Addison Road route was not included. The Walton proposal was included, and it looked like the good people of Walton-le-Dale would soon have their tram service, after 10 years or more of campaigning. It was not to be, however, as the outbreak of World War 1 put the expansion on ice.

Despite the worsening situation brought about by

the hostilities, work was put in hand that year on another extension to the depot, which included new office accommodation on Deepdale Road. This is still in use today by the employee-owned Preston Bus Ltd.

For the duration of the war, the undertaking was greatly aided by the recruitment of women conductors. Economies, needless to say, had to be made, but to an extent nowhere near as drastic as those on the buses during World War 2. For example, the last cars were timed to leave town at 10.30pm instead of 11.10pm and frequencies were reduced. These bit harder towards the Armistice, and in 1918 Sunday services did not begin until 2pm and the last daily departures were brought forward again, to 10pm.

With a return to peacetime conditions, the Tramways Department was faced with expected problems. Neglected trackwork needed attention, the Ashton route being in the worst state of repair, and in 1919 re-laying started. This, of course, provided employment for those men discharged from the armed forces, who viewed the continued use of conductresses with disdain, the tramway men's union voicing strong objections to their retention. It was also realised by many that further work would be available should the expansion programme be reinstated. It never was, and the residents of Plungington had, to some degree, replaced those from Walton as the activists. They, along with others from the Adelphi and Addison Street areas, began exerting pressure on the council to at least secure a bus service, if the tram route was not going to materialise. (Powers had been obtained within the 1914 Act to operate motorbuses.) Their efforts were rewarded in January 1992, with the introduction of the corporation's first bus service, a circular from Fishergate along Fylde Road, Lytham Road, Plungington Road, Friargate and back to Fishergate. Three vehicles were purchased for it, Leyland G7s with English Electric single-deck bodies.

This was also the year Harry Clayton succeeded J. Simpson as Manager. Mr Clayton was to superintend operations until 1946, completing over 40 years' service with the undertaking.

However, back to 1919. Also significant was the augmentation of the tramcar fleet with the addition of nine second-hand single-deckers from Sheffield, who were converting their fleet to all double-deck. (Barrow, as will be recalled, took the other six disposed of.)

The early 1920s witnessed another Guild, in 1922, and, just as at the latest in 1992, the town's transport undertaking needed to be on its toes, as the many processions created havoc with the scheduling of services; a tram cannot, like a bus, be diverted! The obverse of the coin was, nevertheless, the ability of the cars to move vast numbers of people, who thronged into the town for the week.

The decade also saw the birth of a tradition at Deepdale, surviving to this day, in the department's avid 'Do-It-Yourself' policy, starting in 1924 with the fitting of a single-truck car with vestibules and enclosed top, giving the system its first completely enclosed tramcar. A far more ambitious scheme was embarked upon two years later, with the construction of a 'lowbridge' double-decker for the Ashton route under the Fylde Road bridge.

This was in response to the rejection by the Finance Sub-Committee of the council of a proposal to purchase a new car from English Electric. Twelve months previously, the Manager had costed the conversion of an open-top car into 'lowbridge', covered configuration, to reconstruct an ex-Sheffield car or buy new from English Electric or Brush. The 'new' car was, in fact, an ingenious marrying up of one of the original 1904 cars and a Sheffield machine. Two more followed, and once the trio were in service, work started on enclosing most of the remaining fleet; exact numbers are not known.

That mention of Brush is, in itself, interesting, as all the Preston fleet were built in the town by Dick, Kerr and later English Electric, and it is perhaps appropriate at this juncture to break off from the story and take a closer look at that firm.

Strand Road runs from what was the Penwortham horse tram terminus at the foot of Fishergate Hill roughly northwards towards Ashton. At the time of writing the upper half of it is flanked by two long, imposing industrial premises: on the west side, GEC Alsthom and, poignantly, on the eastern side, the empty shell of what had been until late in 1992, British Aerospace, poised for demolition. This latter site is believed to have started life as a railway carriage building works in the decade 1830-1840, later functioning under the auspices of the North of England Railway Carriage & Iron Co until 1878, when it went into liquidation. After being empty for some 20 years, it was bought by the Electric Railway & Tramway Carriage Works Ltd in 1898, who further expanded the site, covering 13 acres.

The company exploited a void in the tramcar supply market, recognising that the already established larger undertakings had to purchase from the USA, due to the incapability of British firms to meet demand. It is obvious that the directors of the Electric Railway & Tramway Carriage Works were astute in anticipating the growth of tramways in this country. These thoughts were echoed by the Equipment Syndicate Ltd of Manchester, who acquired the land to the west side,

Right: Between 1924 and 1928, nine or so of the 1904 cars fitted with top covers, were fully-enclosed, receiving more powerful motors at the same time. No 6, again, is shown after undergoing an attempt at de-enclosing. The 'R' for Ribbleton is worthy of note. *Author's Collection*

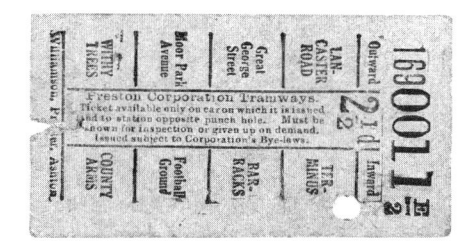

Above: A Preston Corporation Tramways ticket for the Fulwood route. *National Tramway Museum*

Below: Is this the odd-man-out in the Preston fleet, an integral, double-deck, steam tram; the one that got away, a missing link, etc? Sorry, it's just No 18, now sporting a top cover and photographed with a mill chimney behind! The location is the Rose Bud, at the London Road end of New Hall Lane, where, if plans had come to fruition, a line would have diverged off on the right to Walton-le-Dale, in front of the splendid example of a Victorian cast-iron urinal. *Reproduced with kind permission of the Lancashire Evening Post*

Above: Another UEC builder's record shot, this time of Preston's No 33, a single-decker built in 1912, complete with top lights, clerestory and ornate ironwork over the destination box. Unupholstered longitudinal seats were fitted. *Roy Marshall Collection*

Below: By the time this picture of sister No 31 ascending Tulketh Brow *en route* to Ashton was taken, the scroll had gone, as had the 'Preston Corporation Tramways' legend, but most noteworthy is the blanking off of two of the entrances. These three single-deckers were purchased specifically for the Ashton route because of the low bridge already mentioned. *Roy Marshall Collection*

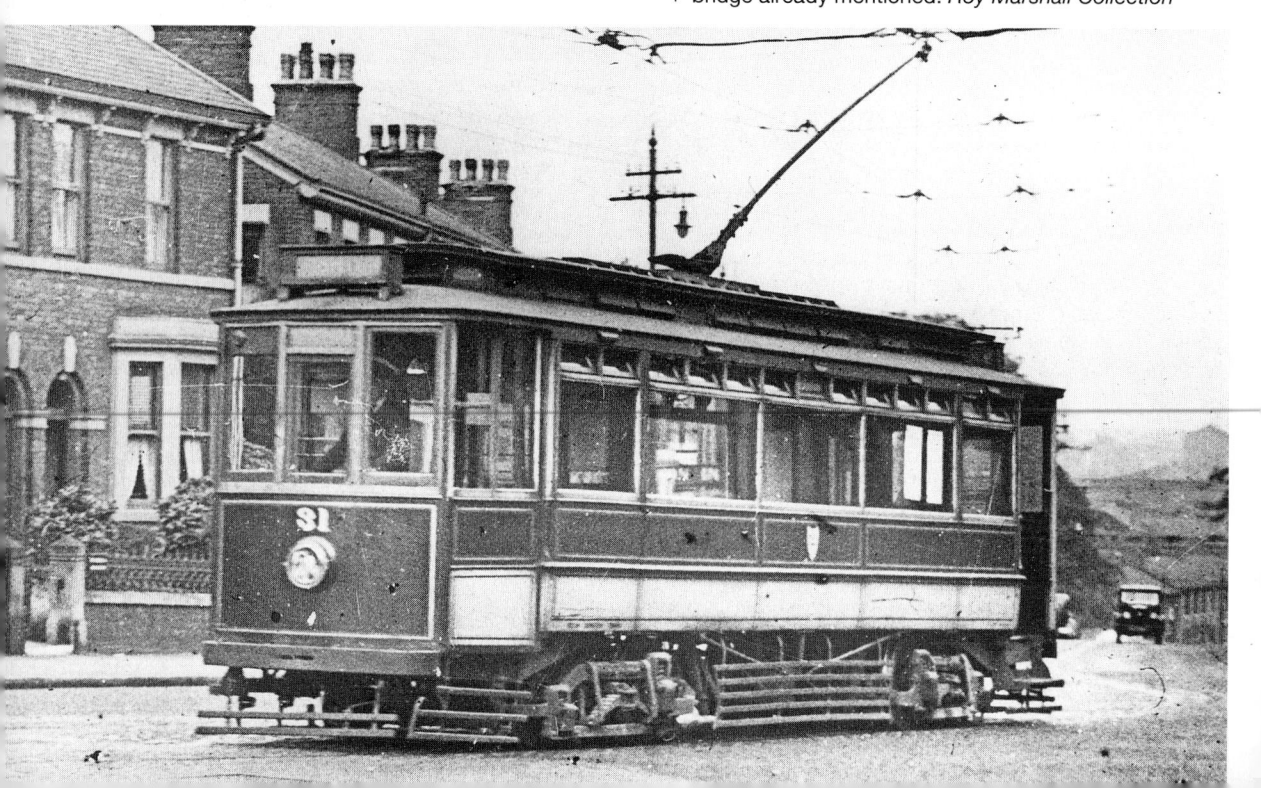

Right: UEC supplied six balcony cars in 1914, mounted on Preston flexible trucks. This is No 39, the last. The retractable sun visor, with more than a passing allusion to shop practice, is visible. *Roy Marshall Collection*

which had been reclaimed from the Ribble, diverted to allow construction of the Albert Edward Dock to begin in 1885. The engineering works were intended for the manufacture of traction and associated equipment. Backers of the Equipment Syndicate Ltd were Dick, Kerr & Co Ltd of Kilmarnock, an old-established engineering firm, noted for its diverse production and supply of complete tramway systems. All this fevered activity occurred around the turn of the century and, by 1900, the Equipment Syndicate had become the English Electric Manufacturing Co, which found itself an associate of the Electric Railway & Tramway Carriage Works' parent company across the road, an effective final amalgamation coming about three years later, when the English Electric Manufacturing Co lost its independence within the group, the whole complex becoming Dick, Kerr & Co Ltd.

Further acquisitions and amalgamations saw such well known names as the Shropshire company, G. F. Milnes, succumbing to the Dick, Kerr-led onslaught, and in 1905 the Preston company name was changed to the United Electric Car Co, leaving only Brush and Hurst, Nelson as the other major independent tramcar builders. In 1918, the English Electric name was resurrected and, thereafter, tramway products were manufactured and sold under that title, as were railway locomotives for home and abroad, along with trolleybuses and bus and coach bodies.

The outbreak of World War 2 in 1939 drastically changed English Electric's operations. Aircraft production soon ousted tram and bus body manufacture, although, interestingly, it was only on the launch of the Canberra bombers and Lightning fighters after the war that English Electric actually built aircraft of its own, wartime output being on a contract basis. The last bus bodies were turned out in 1942, and, although railway work was resumed in 1945 or thereabouts in West Works, East Works remained dedicated to the aircraft industry to the end. Sadly, but inevitably, no more English Electric cars would venture out on to the company's test track, which wound its way around both works and across Strand Road. (A peep into the GEC premises reveals trackwork in the South Yard to this day, though.) In its latter days of production, advanced streamliners for Aberdeen, Darwen and, of course, Blackpool were amongst the revolutionary vehicles to grace its tracks.

Thankfully, 60 years on, it is still possible to ride on English Electric products in revenue-earning service along Blackpool's sea front; ample evidence of the durability of them.

Returning to Preston Corporation and 1926, when its generating plant closed on the opening of the New Ribble Power Station at Penwortham. Current was, after 22 years of being produced 'in-house', to be purchased from the local electricity company, through an AC switchboard and thence two rotary convertors of 500 and 750kW at Deepdale.

On the closure of the Lincoln system in 1929, three of its cars were acquired. Double-deckers, they were, like those from Sheffield, 'coming home', being English Electric-built. Of 'lowbridge' design, they were ideal for the Ashton route. Despite these arrivals, it was obvious the tramway's days were numbered, as in November the same year, following an inspection of the Wolverhampton and Bradford trolleybus systems, it was agreed to seek Parliamentary powers to convert the tramway to 'trackless.' Consequently, the Act came into force on 27 March 1931, permitting the running of trolleybuses over existing tram routes, with an option to extend to Grimsargh. This village, midway between Preston and Longridge, where the Whittingham Hospital Tramway left the Longridge branch, became part of the borough following the 1974 Local Government reorganisation. After deregulation of the bus industry in 1986, Preston Borough Transport, on deciding to operate all its services within its authority area, however, chose not to extend its Ribbleton service to that point. History, of course, does tend to repeat itself, because, as will be seen in a moment, trolleybuses never reached there, either.

All of this came about in the same year as the publication of the far-reaching report of the

Commission on Transport, and the Act stated that the Minister was empowered to order the abandonment of the tramway. Thus, the process had started in earnest and later that year the first discussions took place around a proposal to close the Farringdon Park and Penwortham routes. The erosion continued, and by December, permission had been granted to run buses to the new Greenlands housing estate, further on from the Ribbleton tram terminus at the Bowling Green. Tram frequencies were reduced to allow alternation with bus timings. Minds were now made up, and the title of the department was changed from 'Tramways' to the perhaps by now to be predicted, 'Transport'. On 18 January 1932, the council agreed to convert the 'FP' and 'P' routes to petrol or diesel buses, as the trackwork on these two was in urgent need of renewal. A trial (believed to have been in 1929) using a new English Electric trolleybus for Bradford, which ran without passengers over, probably the Ashton route, near to Strand Road, trailing a skate in the tramline for negative earth, had not been sufficient to convince members to retain electric propulsion. The die was now cast; Preston's days as a tram operator were certainly numbered. Tram services ceased on those two routes on 4 July 1932, although it is believed some peak workings were by tram after the official closure date.

Ribbleton was next to go, and, again, despite the official date in November 1933, part-time working lingered on until 1934, perhaps to about the same time as the final demise of the Ashton route, whose closure had been delayed by certain insistencies of the Traffic Commissioners in respect of the replacement buses. Fylde Road bridge was to continue to create problems into the motorbus era, and eventually 'lowbridge' English Electric-bodied Leyland Titan TD3cs were purchased for the 'A' route. 'Lowbridge' vehicles continued on it until the road lowering already referred to.

By the beginning of 1935, only the Fulwood and Deepdale circulars remained, along with 24 cars to maintain them and the associated football specials, remembered by many older North End supporters for their gross overloading, and the sparks from beneath them as they 'bottomed' on Deepdale station bridge!

They almost saw the year out, being finally withdrawn on 15 December, without any official celebration; the abandonment had cost enough already. All told, in its comparatively brief 31-year history, Preston Corporation Tramways had carried over 370 million passengers and its cars had trundled over 32 million miles.

Left: An engineering tradition. Preston's 'new' No 42 was the result of marrying up parts of one of the 1904 four-wheelers with some of the original No 42, a second-hand single-decker bought from Sheffield in 1918. It emerged from Deepdale in 1928 and two more followed, all to a 'lowbridge' design suitable for the Ashton service, permitting covered top cars to pass under the Fylde Road bridge. This car was later sold to Lytham St Annes (qv). *Author's Collection*

Above: Sister car, No 30, seen at the Ashton terminus. The mirrors are worthy of note. *National Tramway Museum*

Below: With County Hall (built on the site of an early horse tram depot) in the background, a Leyland Tiger bus owned by Singleton's of Leyland is pursued over the railway bridge by car No 7, working across town from Broadgate to Farringdon Park in this early 1930s view, whilst a fully enclosed example of the same batch is about to descend Fishergate Hill. *Author's Collection*

Top: By the time this view of the depot was taken, the tramway abandonment was all but complete. Of the buses evident, one of the replacement 'lowbridge' Leyland TD3cs, just visible through the doorway, confirms that only the Fulwood/Deepdale circular route remained, so the date is probably early 1935. The TD2 on the right was one of the first batch of replacement vehicles, acquired in 1932 to allow the abandonment of the Farringdon Park and Penwortham (Broadgate) routes. Car No 12, on the left, was originally No 48, an ex-Sheffield Brush car, acquired in 1920. *Author's Collection*

Above: A bleak view that sums up so much about the demise of the tramcar. At the back of Preston's Deepdale depot, two of the 1904 cars await the breaker's torch, the one on the left already having lost its top deck. The building in the background is the original power station. A sinister aspect of this is the fact that this is a Leyland Motors photograph. Were they gloating? *Author's Collection*

CHAPTER 6

EAST LANCASHIRE AND ROSSENDALE

This could be considered the heartland of tramway operation in the region, not least by the magnitude of complex activity within it, a state of affairs exacerbating any rationalised study. There are, of course, two sub-regions geographically and, to some extent, economically; Blackburn, Darwen and Accrington being built on the cotton industry, whilst boot and shoe manufacturing was (and still is) prominent in the Rossendale Valley. Therefore, on the face of it, two chapters might be considered obvious, but in-so-far as their tramway histories are concerned, they are inextricably linked.

Of the four companies in the pre-municipal control days, two crossed boundaries, and once they had passed totally into council hands, one corporation undertaking (Accrington) extended into Rossendale. All told, no less than six tramway-owning councils can be counted, one of which never owned a tramcar and another only briefly. On top of this, a municipal operator from outside the region laid tracks into it, and the trolley buses of another, which had intended to build a tramway, also interloped.

Had parochialism not unfortunately prevailed, the area would have seen a daunting 4ft gauge interworked system, extending 16 miles across, and permitting the through journey from South Manchester as referred to in Chapter One, 20 odd miles of it within East Lancashire and Rossendale. In fact, it is reckoned that only on one occasion, in June 1900, to mark the end of steam traction in Darwen, did a tram venture from that town's

Below: The Blackburn & Over Darwen Tramway Co shows off its new tram engine No 8, and trailer, in 1885. This was the precursor of seven engines from Thomas Green to supplement a similar number supplied by Kitsons for the opening of the line between Blackburn and Darwen in 1881. This route was the first in the country to be authorised for steam. Passenger comforts were not overlooked over a century ago; curtains in the saloon and a half canopy to protect the unfortunates on the top-deck from the smoke and sparks belched out of the engine's chimney. Altogether, there were 23 trailers. This one was single-ended, employing the Eades patent. *Author's Collection*

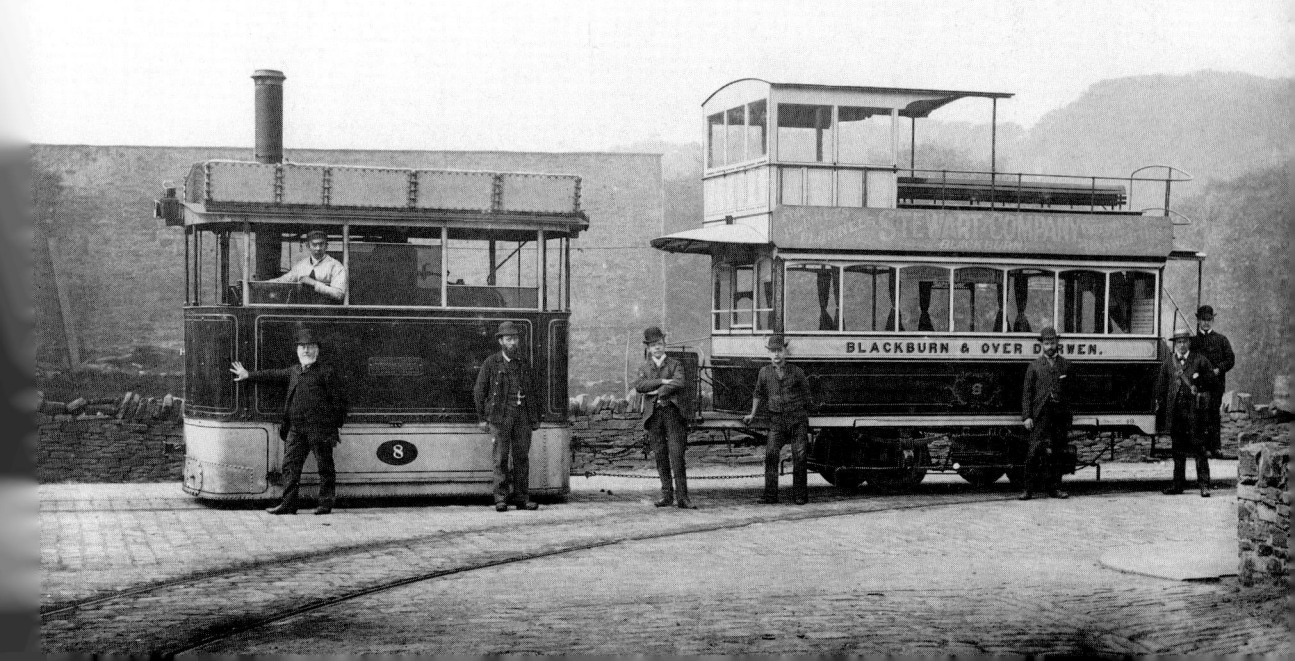

Whitehall terminus, through Blackburn, Accrington, Haslingden, and Rawtenstall to Bacup, 21 miles away, with a civic party on board.

Back, however, into the last century, to start the unravelling process, by looking at those four companies. First on the scene was the Blackburn & Over Darwen Tramways Co, which, having obtained Parliamentary Powers in 1879, commenced working the five miles between Blackburn and Darwen on 13 April 1881, using, initially, seven Kitson locomotives. The line was single throughout, with a depot in Lorne Street, Darwen. Next came the Accrington Corporation Steam Tramways Co, five years later. This was a joint venture between the council and a private concern, which leased the tracks and other infrastructure. Work had started on the nine-mile route in 1884, being ready for opening on 5 April 1886. Here we have the first inter-sub-regional working, with the original three routes radiating from Accrington to Church, Clayton and Baxenden being supplemented by its Haslingden and Rawtenstall extension, completed as far as Lockgate on the Haslingden/Rawtenstall boundary in August 1887.

Keeping things in chronological order, Blackburn saw the third company manifest itself in the form of the Blackburn Tramways Co, owned by a London syndicate. It had obtained powers in 1885 to open four routes in the town, and the first one was inaugurated on 28 May 1887, to meet up with Accrington's cars at Church. The second followed in January 1888, to Wilpshire. Both were steam-hauled. The remainder of the eight-mile system (all of it leased from the corporation) used horse trams on the lines to Billinge End and Witton Stocks, opening in 1888 and 1889 respectively. The company amassed a fleet of 14 engines, 19 trailers and eight horse cars (requiring 70 horses). There were two depots: the first, for the horse trams, was in Simmons Street, near Sudell Cross, and still stands, as does the other, for the steam fleet, now greatly expanded as Blackburn Transport's bus garage at Intack.

The final player was the Rossendale Valley Tramways Co, which opened a steam-worked route from Rawtenstall to Bacup in January 1889, soon to be extended to Lockgate to meet up with Accrington's. Another line, northwards from Rawtenstall to Crawshawbooth, in the Burnley direction, was completed in 1891, giving the company 6.35 miles of tramway. This would have been increased to all but 10 miles had the authorised section from Bacup to Facit, between Whitworth and Rochdale, which formed part of the latter's later electrified line, materialised. Thomas Green engines were the first choice, 11 having been taken into stock by 1894, with an ex-Blackburn Tramways Co locomotive following, on that company's closure in 1901. In 1900 the company was purchased by the British Electric Traction Co, with obvious intentions to electrify but, as will be seen, they were overtaken by events.

So far so good; things seem reasonably straightforward, but the municipalisation of the tramways in East Lancashire and Rossendale will make the complexities referred to in respect of Morecambe look crystal clear. Therefore, irrespective of the several common threads, each of the evolved corporation undertakings will be described alphabetically.

ACCRINGTON

The corporation, of course, owned the permanent way on the three routes within the borough and the depot in Ellison Street, but the trackwork on the Haslingden and Rawtenstall extension was owned by those two corporations. So, on exercising its Tramways Act options by buying the company in September 1907, the line into the Valley was not to be part of the £2,227 deal. However, as will soon become apparent, that was not to be the end of Accrington cars 'over the hill'.

All this had been made possible by the Accrington Corporation Act of 1905, which also authorised two new routes, the first being an extension of the Church line as far as New Lane, Oswaldtwistle. Such an extension had been impossible in steam days because of the limited clearance under the railway bridge at Church & Oswaldtwistle station, overcome by the intended use of single-deck cars. The other was along Burnley Road to the cemetery at Hillock Vale. A short extension of the Clayton-le-Moors line of about 200yd from the 'Load of Mischief' public house to the canal bridge was also authorised. (As an aside, 'Load of Mischief' was often appropriately displayed in later years on corporation buses hired by Sunday Schools etc, for trips to such places as Blackpool.)

The Rawtenstall line could only be electrified by the corporation as far as the Haslingden boundary at Baxenden, as the lease on the Haslingden section had yet to expire. It was, of course, that council's prerogative to make the necessary arrangements for take-over and electrification. To progress matters, a sub-committee was formed to research and make recommendations as to what was needed for the proposed system, and consequently, the Electricity and Tramways Committee decided that 14 double-deck and four single-deck cars would be needed, the tender from Brush being accepted in December 1906, by which time Hadfields of Sheffield had been awarded the contract to supply pointwork,

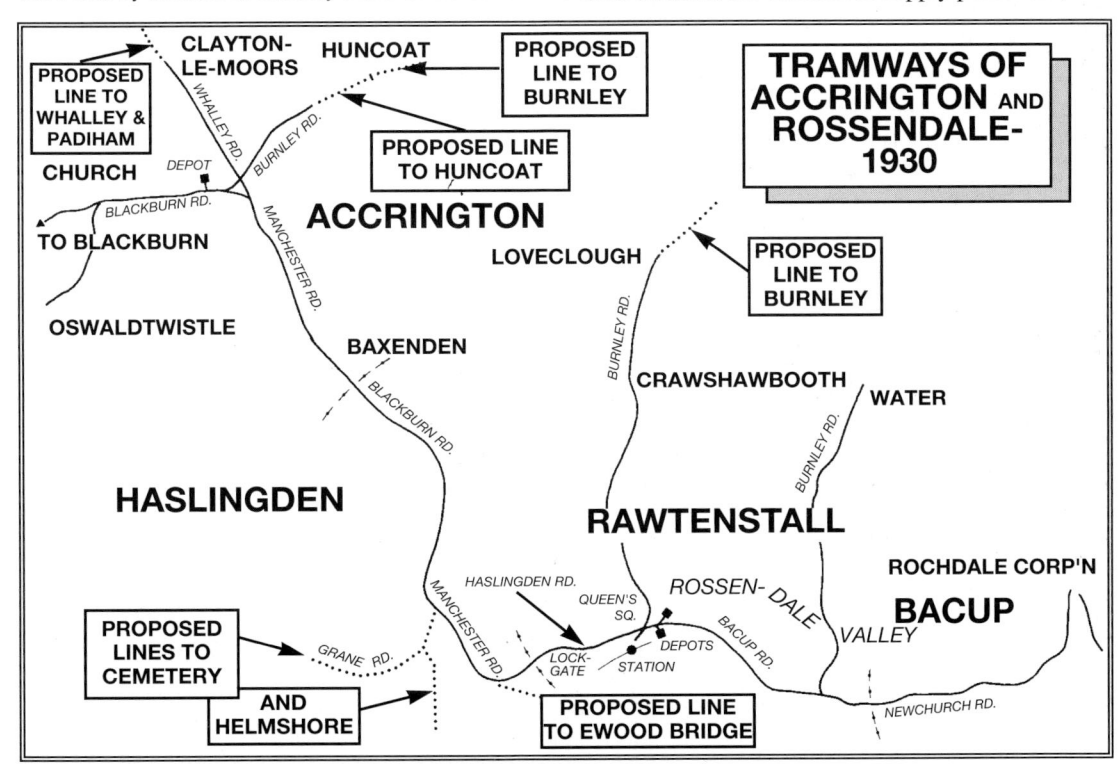

Above left: Ceremony on the Accrington Corporation steam tramways system, probably at the Haslingden boundary, as work was to commence on the Haslingden and Rawtenstall extension, completed as far as Lockgate in 1887. All its engines — 23 including four purchased from Blackburn Corporation in 1901 — were built by Thomas Green. *Roy Marshall Collection*

Above: Blackburn Corporation Tramways Co Ltd employed both horse and steam traction. No 10 was a Thomas Green engine bought for the inception of the two steam routes, to Church and Wilpshire, and is seen here coupled to trailer No 15, built by Falcon in 1888, on the latter. Salford Bridge was the town centre terminus and the cemetery the intermediate point in the journey. *National Tramway Museum*

crossings, etc, with that for the 1,500 tons of rail and 25 tons of fishplates going to Walter Scotts of Leeds. Dick, Kerr was pipped at the post by Brush for the overhead part of the job. Naturally, the corporation's power station would be unable to cope with the increased demand of the system, and a new boiler was ordered, along with two Bellis and Morcom engines, a pair of 50kW dynamos, a generator set and switchgear.

Work was progressing on the electrification, but it was obvious as the new year of 1907 dawned, that it would not be ready in time for the take-over on 7 April, when the company's lease expired. It was therefore agreed that until everything was completed, the tramway would be operated jointly, with the company retaining ⅜ of receipts, passing the remainder on to the council.

That was the easy part, as neither side could agree on the terms of the take-over, especially with regard to the value of the fleet. An arbiter, Mr Reginald Page Wilson, was called in, and on 31 July, he announced his decision: £2,255 for the engines and cars, if the settlement was to be on a landlord and tenant basis, or £880 if the rolling stock was deemed to be of scrap value only, irrespective of a further £235 for sundry materials. On 20 September, agreement was reached on the £2,227 figure mentioned earlier. The corporation had already accepted an offer of £1,230 from the Sheffield scrap merchants, T. W. Ward & Co for the engines and cars and 63s (£3.15) per ton for the old rail; a nice profit!

The proposed opening date was to be 2 August for the Oswaldtwistle section, with the rest to follow, but by early July this was looking unlikely. Although the permanent way was on schedule, Brush had to be taken to task for their slow progress on the overhead, compounded a little later by complaints from the corporation about the paintwork on the standards and deviation from the drawings in respect of some pull-off span wires.

Accrington was not having an easy time with its project, some of it being brought upon itself by poor planning, the omission of estimating for uniforms being typical. Meanwhile, Brush, no doubt anxious to get something right in the eyes of the council, began delivering the cars. All well and good, except for the fact there was nowhere to put them, as the shed extensions were not ready, necessitating temporary track having to be laid to

Top: Oh calamity! A shambolic state of affairs on the electrification of the Accrington system, with new cars arriving from Brush and the depot far from complete. *National Tramway Museum*

Above: Of the 18 cars bought from Brush in 1907 to commence operation, four were 32-seat single-deckers. Here, No 4 comes under the scrutiny of officials. High-mounted large headlamps remained popular with Accrington to the end. *National Tramway Museum*

Left: The rest were balcony cars, again by Brush (as were all the fleet). Here, No 11 is seen working through to Haslingden. On crossing the boundary, it would have traversed Haslingden Corporation metals, over which Rawtenstall cars also ran. *Roy Marshall Collection*

Below: Haslingden did have a flirtation with tramway operation, albeit for the brief period following the joint acquisition of the Accrington steam company, when it ran eight trams pending electrification, and leasing of its system. This is the change-over from Accrington electric tram to Haslingden steam at Baxenden station. *National Tramway Museum*

free up the open air parking, occupied by already withdrawn steam fleet vehicles.

Much to the disappointment of the doubters, the target date for the Oswaldtwistle opening was met, Board of Trade approval being granted on 1 August, and with the usual civic pomp the first cars ran the next day. Another feature of the line was a double junction at the old Church terminus, creating a physical connection with Blackburn's tracks. With the opening of the new line up to Baxenden on 1 January 1908, what could be completed, pending negotiations with Haslingden, had been. Most of the system was single track with passing loops, double sections being employed on some bends, where for safety reasons the next loop would be out of sight of the motorman. However, on Union Road, Oswaldtwistle, which was narrow, this was not possible, so automatic signals were eventually installed to warn of an occupied section ahead, this being prompted by a couple of collisions during the first months.

Accrington Corporation cars reached Lockgate on 20 October 1908, on Haslingden's completion of the electrification of its system. Accrington now had its eyes on exploiting further potential, and on 4 May 1909, approached Rawtenstall Corporation, the new owners of the electrified Rossendale Valley line, with a view to joint through running to Bacup. Rawtenstall was not happy with this but acceded to through running as far as Queen's Square, Rawtenstall. However, Bacup was reached on 1 April 1910, after considerable wrangling over the cost of power, which became complicated, to say the least, the three corporations being initially unable to agree on aggregate amounts consumed.

Back in town, it soon became apparent that the Burnley Road route was not going to be profitable and, as early as September 1908, double-deckers were substituted by single-deckers, at a reduced frequency. An interesting feature of this line was the provision of special funeral cars up to the cemetery, for years a bone of contention by the disgruntled Carriage Proprietors' Association. The tramway, thus, settled down to more or less routine operation. World War 1, with its problems, came and, thankfully, went, women conductors and a handful of drivers being recruited to keep the wheels turning, and from 1917, through running into the Valley was curtailed at Queen's Square, never to be reinstated in tram days. This was not through acrimony, though, as the good relationships between the five corporations was exemplified that summer by the joint purchasing of second-hand rail and overhead from a source that remains something of a mystery to this day.

Below: In 1917, No 9 was one of three cars (Nos 10 and 17 were the others), that were converted to open-top. It is depicted here in Accrington Market Place. *National Tramway Museum/Whitcombe Collection*

Above: Two cars arrived in 1912, a balcony and a single-decker. This is the former, No 26. *National Tramway Museum/Whitcombe Collection*

Below: In 1915, to cope with increasing demands, especially on rush-hour workings on the Church and Clayton sections, five new cars were ordered from Brush, but the outbreak of war delayed their delivery until 1919. Three eventually materialised as 40-seat single-deckers with clerestory roofs, as depicted by No 28. Bogies appeared in the fleet for the first time with these cars. (The balance of the order was two double deckers, including No 39, illustrated in Chapter One.) *Roy Marshall Collection*

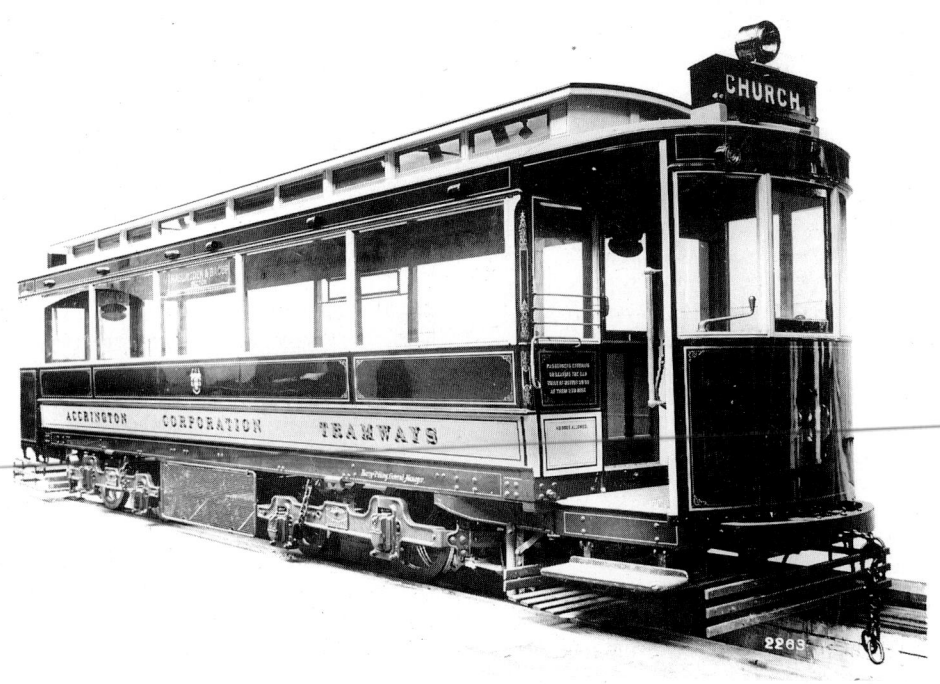

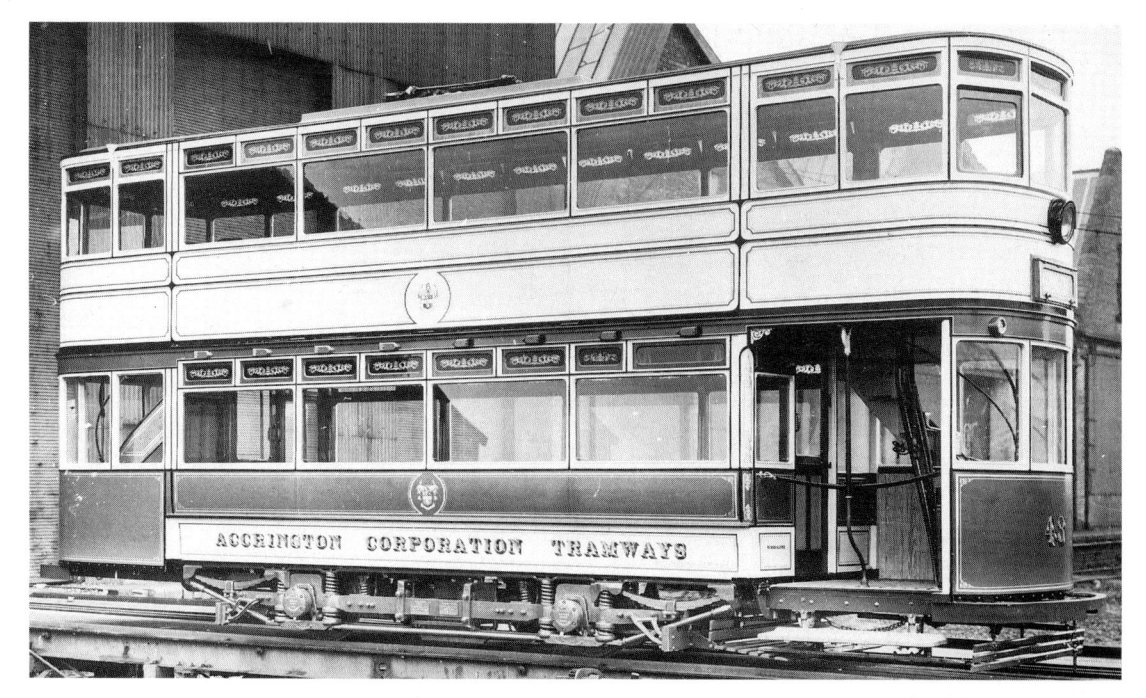

Normality following the Armistice was not to reign long, for in 1922 Ribble secured a licence to run between Accrington and Burnley, with conditions restricting the picking up and setting down of passengers within the two boroughs. At the same time the Rishton & Antley Motor Co also obtained a licence, but for a service within the borough, to Green Howarth, an isolated part of town. It could pick up where it wanted, as it did not compete with a tram route. Despite the obvious, Accrington Council 'slurred its clogs', to use a local expression, and by the time it applied to run buses itself, the law had been altered, requiring municipalities to obtain Parliamentary powers to do so. This was done and granted under the Accrington Corporation Act of 1928, giving the council authorisation to operate buses or trolleybuses within a five-mile radius of the town hall. The now familiar pattern emerges, with Accrington's first services, to Huncoat, Willows Lane and Woodnook beginning on 12 November 1928. In 1930, along with Ribble, the corporation acquired Rishton & Antley, providing joint services to Green Howarth, Whalley, Clitheroe and Burnley. The inevitable ironies figure here: powers had been obtained in 1901 by the Blackburn, Accrington, Padiham & Whalley Tramways Co to link Padiham with the Clayton-le-Moors terminus and by the Accrington & Burnley Light Railway to provide a connection at the cemetery extremity.

Above: For the final cars purchased by Accrington, bogies gave way to four-wheel trucks once more and the pair, received in 1926, were of special 'lowbridge' construction for the Oswaldtwistle service. This was achieved with the use of the Brush patented well-construction for the floor. The seating arrangement was similar to that on the earlier bogie cars. *Roy Marshall Collection*

Nothing had come of the proposals, which in one respect would have rejuvenated the loss-making Burnley Road route. Whilst that line never returned a profit, the opposite was the case with regard to the Oswaldtwistle route which, in 1926, saw the introduction of special low-height double-deckers to augment the single-deckers. These two were the last cars delivered to Accrington and were built by Brush, to whom the council had remained totally loyal.

Motorbuses were by 1930 gaining a foothold, and in March the Accrington-Rawtenstall line was closed, with buses intended to take over the Oswaldtwistle service on 1 August 1931. What was left was deemed unsustainable, and therefore the Clayton line was brought into the abandonment programme, closing on 26 August, 1931, along with Oswaldtwistle, which had lingered on for three weeks. With yet more irony, it was to be the Burnley Road route that saw the last car on 6 January 1932.

BLACKBURN

The distinction of having the first electric tramway in this part of Lancashire goes to Blackburn, which could boast another claim to fame through its extensive use of interlaced track, believed to be, pro-rata, the most in the country. As will be remembered, two companies operated in the town, one of which worked into neighbouring Darwen. Foolishly, in 1884, Blackburn Corporation had told the Blackburn & Over Darwen Co that it had no intention of buying it out until 1914. However, realising their mistake, councillors suggested to their opposite numbers in Darwen, in 1898, that they should buy the section of the line in their borough, making the company's position untenable. In turn that would make it amenable to a Blackburn take-over, thus saving the corporation's face. This bit of double-dealing paid off, Blackburn paying a total of £108,620 for its share of the B&OD company and the outright purchase of the Blackburn Tramways Co.

The horse tramway routes were the first to be electrified, the 4ft gauge track being completely replaced. Amazingly, the conversion of the two lines, along Preston New Road to Billinge End and to Cherry Tree, an extension from the original horse tram terminus at Witton Stocks, which had, incidentally, been subject of an approval in company days for a shorter extension to Fenniscliffe Bridge, was completed by 20 March 1899. Eight 60-seat open-top cars were supplied by Milnes for the opening. Blackburn was always a 'big' car user, and these inaugural ones were no exception. Mounted on US-manufactured Brill bogies, they also set the future standard in-so-far as chassis were concerned. The 61 cars acquired over the years were all to be mounted on bogies either of this make or, in the case of the next large batch of 40, also built by Milnes, by Peckham. Received in 1901, they were set to work on the rest of the rejuvenated system, firstly the former B&OD line to Darwen, jointly worked with that corporation until 1946. Next to open was the route out to Church, known as the 'Country Line', for in those days the area between Blackburn and Accrington was mainly of a rural nature, offering exhilarating and often high-speed runs through the fields, perhaps rivalling those along the Blackpool & Fleetwood's cliff tops!

Two miles out of the town centre on this line was the old steam depot at Intack, the line to which had been electrified some time before Church had been reached on 9 January 1901. Initially, the new electric fleet had been housed at Simmonds Street, a somewhat cramped site, so the more spacious surroundings at Intack leant themselves to possible expansion, the deciding factor when a permanent and larger home for the growing number of cars was sought. Wilpshire, on the road out to Whalley, was reached on 14 May 1902, when the steam line to the cemetery along Whalley New Road was extended to the borough boundary and, of course, electrified.

There was, however, more to come. Although the existing lines were now upgraded and electrified, a new one to Queen's Park, via Audley Range, known as the Audley section (Blackburn's routes were always referred to as such) saw its first cars in 1903, bringing the total mileage to 14.73. The modernisation process had cost the council £150,000 on top of the purchase price, which made quite a dent in the Borough Treasurer's pocket. So much so that a loss of £34,000 was recorded in the first year. This was, thankfully, soon turned round, reflecting the wise investment and shrewd (if not, at times, devious) dealing by the corporation.

Besides the previously referred to extensive use of interlaced tracks, the Blackburn town centre layout was interesting. Although not unique, the streets there were narrow and certainly not ideal for double trackwork. Therefore it was decided that lines entering the Boulevard terminus, in front of the railway station, would leave by a different route, creating a complex, but very practical, one-way system. The Wilpshire route, however, did not terminate in the Boulevard, but rather within a short 'Y' junction at Salford Bridge, and on its outward journey ran parallel with the Church line for a short distance. The accompanying diagram amply illustrates this fascinating facet of the system.

At Ewood Park, a mile or two from the Darwen boundary, is the home of Blackburn Rovers

Right: The joint working between Blackburn and Darwen from the beginning of the company days was perpetuated after electrification, and Blackburn Corporation's No 29, the second of the new electric cars, received in 1899, is seen here at Darwen Circus. Interestingly, the numbering of the cars was carried on from the company's series. Milnes was the builder of these initial eight cars, which, as with all subsequent deliveries, were mounted on bogies, in this instance, Brill 22Es. The off-set trolley pole, as seen in Blackpool, was also employed here. *A. D. Packer Collection*

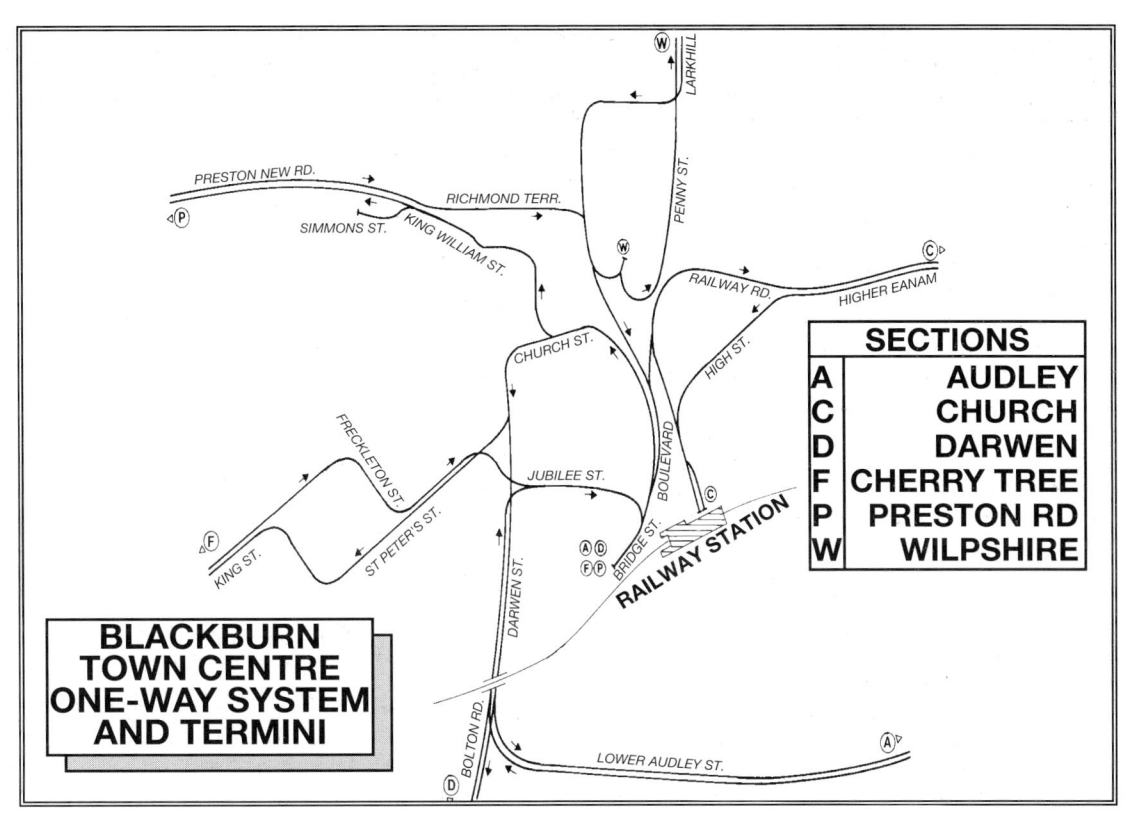

BLACKBURN TOWN CENTRE ONE-WAY SYSTEM AND TERMINI

SECTIONS	
A	AUDLEY
C	CHURCH
D	DARWEN
F	CHERRY TREE
P	PRESTON RD
W	WILPSHIRE

Left: A large batch of 40 cars was purchased in 1901, again from the Milnes stable, which seated 73. Resplendently lined out and in dark sage green and cream livery, No 54 is seen at the undertaking's Intack depot, showing 'Audley Section' on its blind. Blackburn always used this expression for its routes. Angular ends became a well-known feature of Blackburn's cars. *Blackburn Transport*

Left: A different style of conversion is shown here on this unidentified car at Intack. Full-drop windows were fitted to the upper-deck and a simplified livery, which was not adopted, was applied. *Blackburn Transport*

Below right: All but eight of the 1901 cars received top covers between 1907 and 1935. Here is No 52, on the Cherry Tree Section, just before World War 2. *National Tramway Museum*

Right: A motley crew! Intack Depot staff in the early part of the century, with an original car flanked by two of the 1901 delivery; in two years designs had improved tremendously. The 1899 cars were known to staff as the Siemens cars on account of their motors. General Electric was to provide units for the rest of the fleet. *Blackburn Transport*

Football Club. Not just in Blackburn, but in effect all over the land, the tramcar proved invaluable for moving the crowds of supporters, and the line in Bolton Road outside the ground soon gained a third track to cater for the specials from both towns. Cars 'laying over' during the game were also accommodated on a spur at right angles to Bolton Road in Kidder Street.

Twelve rather unusual closed combination cars were bought in 1905. These were considered more appropriate than high capacity double-deckers on the Audley section, on which loadings had not come up to expectations. Another unlikely car appeared the following year, destined, though, to be nowhere near as useful as those just mentioned. For some strange reason (perhaps the chairman of the Tramways Committee had taken one of those exhilarating rides to Fleetwood) a toastrack car was built at Intack and put to work on the Wilpshire section. Nobody, understandably, liked it, even on the country route to Church, and it saw very little service. It lay derelict at the depot for many years but was not scrapped until 1937.

The double junction at Church, put in by Accrington, enabled through running between there and Blackburn as from August 1907, which was to be the exclusive preserve of Blackburn Corporation until 1917 when Accrington shared the workings. This was in response to a request by Blackburn for an increase in mileage allowance over the other's tracks, and, after negotiations, it was agreed that adjustments could be made between each of them on a basis of equal mileage being run by each over the whole route. On Accrington's abandonment in 1931, Blackburn cars terminated at Church, passengers having to change to an Accrington Corporation bus for the remainder of the journey.

Despite the appearance of the first motorbuses in 1929, the tramway remained intact until 1935, when the Audley section, not surprisingly, was closed, on 14 February, and plans were prepared for a speedy conversion of the rest. Irrespective of the high esteem in which the system was held, the corporation saw the motorbus as the way forward. A joint service to Darwen, avoiding the tram route, had proved popular and the local independent, the Blackburn Bus Co, which operated to Green Lane, Moorgate and Tockholes, was acquired in 1931.

The outbreak of war in 1939 put the plans on an enforced hold, and during the dark days that followed, the trams earned their keep many times over, moving the vast army of war effort workers. There was to be no gratitude shown to the cars as, in January 1946, on receipt of the first of a large fleet of Leyland Titan PD1 and Guy Arab II double-deckers, the Preston New Road section closed. The demise of the remnants was rapid: Wilpshire on 21 December 1947; Cherry Tree and Church on 16 January 1949; and what was left of the Darwen line after the closure of that system in 1946, curtailing Blackburn cars at the boundary, on 3 September 1949.

In its earlier days, there had been plans to extend the system, approval having been obtained for three new lines: to Revidge from Billinge End, to join the Audley section to Accrington Road and for a line along Burnley Road to the Rishton boundary. In the event, the system was deemed adequate.

Above: After the open-toppers of 1901, came 12 40-seat closed combination cars in 1907 and 1908. Built by UEC, they were originally open-ended, later being fitted with vestibules as seen here. *Blackburn Transport*

Below: Two of the 1908 combination cars are seen in original form in the depot, amidst a selection of Peckham bogies. These cars were intended for the Audley section, where loadings had not come up to expectations. *Blackburn Transport*

Below: A Blackburn Corporation Tramways ticket. *National Tramway Museum*

Above: An unlikely car took to the rails in 1908 in the shape of a cross-bench car No 88. Built in the corporation workshops at Intack on Brill 22E bogies, it was really more suited to Blackpool, Fleetwood or St Annes, and saw little use. *National Tramway Museum*

Below: A 1920s shot of the Boulevard, taken from the railway station. Nearest the camera, adjacent to a line of taxis, is single-deck car No 82. As mentioned in the text, the layout in Blackburn town centre was complex and this part of it will be made clearer when studied in conjunction with the diagram. *Blackburn Transport*

Above: Another 1920s view, this time of the yard at Intack with the department's Vulcan lorry and, behind it, car No 30 of the original 1899 delivery, now with end canopies. This work was done between 1920 and 1923, when their upper-deck capacities were increased from 30 to 42. Most of them eventually received BTH or EE equipment in place of their Siemens.
Blackburn Transport

Below: Behind the scenes work-horse, sweeper/water car No 1, built by Hurst Nelson in 1900 on Brill bogies. However, it was not destined to have a long life in this form, as its chassis was used in 1908 for the cross-bench car, No 88, but its superstructure reappeared in 1914, mounted on an ex-Burnley Brill truck. It is seen here in that condition, centrepiece of a cameo worthy of lingering over, as a fine example of the unseen aspect of tramway working. *Blackburn Transport*

DARWEN

By now the origins of Darwen Corporation Tramways and its close association with Blackburn should be well established. The 'Over Darwen' in the steam company's title stems from the fact that there grew up, during the last century, a Lower Darwen, which became part of the pre-1974 County Borough of Blackburn, and Over Darwen, which is now simply Darwen. The company had its headquarters in Over Darwen, on the site which in 1899 became the depot of the council-owned system on the corporation's exercising of its Tramways Act options, in collaboration with Blackburn. As will be recalled, Darwen was persuaded to acquire the southern portion of the line, within its boundaries.

Upgrading and electrification were the corporation's first priorities, over what was really a very straightforward and limited system; one line, some 3.5 miles in length, between the Blackburn boundary at Earcroft in the north, and Whitehall, at its southern end, on the road to Bolton. The town centre around Darwen Circus lay between.

Most of the route was relaid with double track and was ready by 17 October 1900, although it was to be 1 December before the official opening ceremony took place. Ten double-deck cars, built by Milnes, seating 72 and mounted on Brill bogies, inaugurated the service, and were bedecked in a vermilion and purple lake livery.

The Lancashire & Yorkshire Railway had, in 1876, constructed a line to the village of Hoddlesden, approximately a mile to the east of Darwen, some 250ft up on the moors. It took a circuitous route, leaving the Blackburn-Bolton line at Hollins, north of Darwen station, doubling back on itself to reach the terminus. It never saw passengers, remaining open until 1950 as a freight only branch. Despite the obvious lack of patronage of the railway, the corporation immediately on opening its 'main line' embarked upon building a direct link with Hoddlesden, up the steep Marsh House Lane, and for the last quarter mile or so, along Hoddlesden Road, laid specially for the tramway. This new line was single throughout, apart from the short section between Darwen station and Bolton Road. It was opened on 11 October 1901, using four-wheel cars, again by Milnes. Descending Marsh House Lane was at times precarious and, as a result, in 1905 a small 22-seat 'demi-car' was acquired for evaluation; two more followed in 1906.

Due south of the Whitehall terminus, about five miles away over Belmont Moor, was the terminus of Bolton Corporation's Dunscar route, and in 1901 the Bolton, Turton & Darwen Light Railway applied for powers to link up the two systems. In common with so many other schemes, it failed.

Another line, a loop to serve the village of Sough, intended to leave Bolton Road at the Circus and rejoin it at Whitehall, came to nothing, no doubt the poor receipts on the Hoddlesden branch dissuading the council from risking it. In fact, the Hoddlesden route was proving impractical, and in April 1926, the driver of No 11 of the first four-wheelers lost control in wet conditions and left the rails at the junction with Sudell Road, crashing into a billiard hall. This was not the only accident on this part of the system, prompting its demise, although surprisingly as late as 1937, when it was replaced by buses. A year before, however, it was clear that Darwen saw a future in its trams, as two modern, Luff-style English Electric streamlined double-deckers were bought. There is no suggestion that, like its neighbour, there should have been any plans for prewar abandonment, but the war took its toll on the cars and the permanent way. In 1945 there was another serious accident, which was the deciding factor.

The final Darwen car ran on 5 October 1946, by which time there were only eight left in stock, now painted vermilion and cream, a colour scheme perpetuated until 1974 by the town's equally smart bus fleet, now itself just a memory.

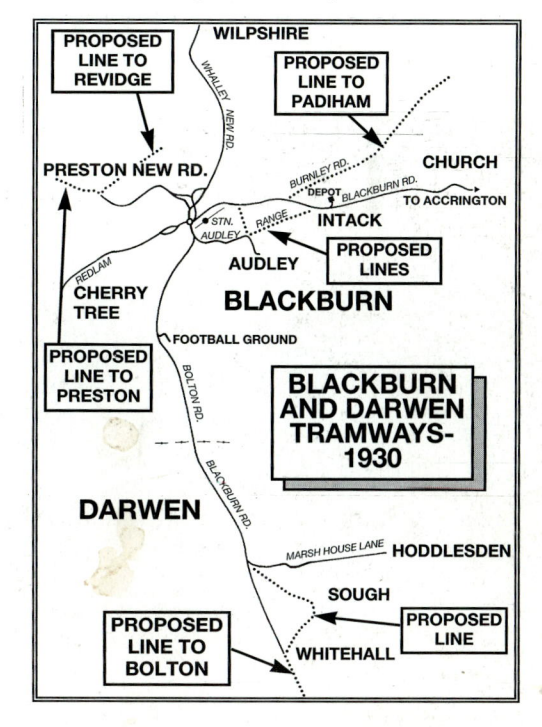

Top: Darwen's No 7, one of 10 Milnes 72-seat open-top cars on Brill 22E bogies, purchased for electrification in 1900. The livery was originally vermilion and purple lake, not the brightest of combinations. *National Tramway Museum*

Above: The following year, four more cars appeared, but this time the driver was afforded the luxury of an enclosed platform, no doubt much appreciated during northeast Lancashire winters! This is No 14. *National Tramway Museum*

Top: Classic tramcar and classic bus pictured together on the Boulevard during the war. Built by English Electric in 1936, No 23 was one of a pair of Blackpool-inspired streamliners. However, the narrower 4ft Darwen gauge somewhat detracted from the intention. Nevertheless, they were smart trams, accentuated by the later vermilion and cream colours. Both passed to the Llandudno & Colwyn Bay Electric Railway on closure in 1946. The Blackburn Corporation bus is No 44, an all-Leyland TD5, new in 1938. *National Tramway Museum*

Above: Second generation No 7 in the Darwen fleet was this fully enclosed car built by the corporation, using new Brush top covers and parts of the lower saloon of the original car. Between 1925 and 1933 six cars were so constructed, although the last (No 10) had utilised an ex-Rawtenstall cover. The location is the Boulevard in Blackburn. *National Tramway Museum*

Darwen tickets.
National Tramway Museum

Left & below: Darwen Corporation leaflets.
National Tramway Museum

HASLINGDEN

Whilst Haslingden shared with Bacup the distinction of owning a tramway but not running cars in its own right, it was, however, an enthusiastic municipal transport operator, albeit of motorbuses. That said, the statement is not totally true. On 1 January 1908, Haslingden Corporation purchased 2.9 miles of line from the Accrington Corporation Steam Tramways Co, along with eight Thomas Green steam engines and seven cars, working the line until 4 September that year. So, briefly, Haslingden was a tramcar operator. Quaintly, one locomotive was retained, and lasted into the 1930s as a snowplough. The corporation did, nevertheless, have ambitious tramway plans, for in 1906 Parliamentary permission had been granted for the construction of three branch lines, totalling 3.25 miles, to the cemetery on Grane Road, Ewood Bridge and Helmshore.

In the event, only the section between Baxenden and Lockgate was electrified and worked by Accrington, as part of the through route to Bacup. Haslingden paid Accrington for the electricity used within the borough and, likewise, a sum for them operating their cars. Understandably, Haslingden supplied its own inspectors.

Accrington rented part of the depot in John Street for £50 per annum to stable four cars, but the arrangement came to an acrimonious end in 1916, when it was discovered that Haslingden had also leased the depot to a haulage contractor, who was storing waste paper there, a state of affairs deemed unsatisfactory and probably dangerous.

Below: All that epitomises the Rossendale Valley surrounds Rawtenstall's No 13 in its early days in pursuit of a pair of haycarts along Haslingden Road towards the town centre.
National Tramway Museum

RAWTENSTALL

Here could have been the knub of a network within a network, with connecting lines to such places as Burnley, Bolton and Bury, had plans been taken through to their hoped for conclusions. Rawtenstall, nevertheless, could boast a respectable 11.75 miles; 4ft gauge, of course, with three branch lines (one, albeit, only 120yd in length) and through running along the 'main line' between Bacup and Accrington.

Electrification of the Rossendale Valley Tramways system came comparatively late, caused by the need for the company's lease to be allowed to expire in 1909, although work actually started on the overhead the previous year. Dick, Kerr had been awarded the contract for the whole job: permanent way, electrical installations and cars. As has been already mentioned, the company worked one branch line up to Crawshawbooth from Queen's Square and Rawtenstall, and within the corporation's powers was authorisation for an extension of this line to Loveclough, 0.75mile towards Burnley, along with two new routes. The first, was to be 2.75 miles up the Whitewell Valley, northwards from Waterfoot, between Rawtenstall and Bacup, to Water; whilst the second was the short branch from Queen's Square to the railway station. The section within the Borough of Bacup was to remain the property of that authority, but be worked entirely by Rawtenstall.

Although the Bacup-Lockgate section was opened on 23 July 1909, it was to be 1 April 1910 before the three councils could agree on the conditions for the whole length. By this time, the first electric cars had reached Loveclough, the section opening on 15 May 1909. From here a connecting line with that of Burnley Corporation at its Summit terminus, a couple of miles away, was projected, but never started. Had it materialised, it would have saved the legs of thousands of Burnley Football Club supporters, who, over the years, trudged between Loveclough and the Summit, and back. The more affluent amongst them, though, would have sped by them on the horse bus service, provided by a Mr Hargreaves to bridge the gap.

The Waterfoot-Water line opened on 21 January 1911. Accommodation in the old steam depot in Bacup Road was augmented by a new building across the road, providing sufficient cover for the initial fleet of 16 balcony cars, all of which were fitted with regenerative braking. This equipment was destined to be shortlived, however. In 1910, a car descending from Loveclough ran out of control

Right: Rawtenstall's No 12 pictured in original form, as delivered in 1909. One of 16 purchased to start electrified operations, it was a UEC product seating 51. *National Tramway Museum*

and crashed, putting in doubt the safety of this type of brake, which made use of shunt-wound traction motors. When running downhill, with the current shut off, the motors acted as generators, returning electricity to the overhead. It worked best when one car was going down, and another up, but it was not always possible to balance services this way, resulting in blown circuit-breakers, burned-out lamps, etc. In the Rawtenstall accident, though, the cause was found to be the failure of a shunt-winding, causing the motors to race, and the Board of Trade at the subsequent inquiry, criticised them, and regenerative braking in general, resulting in cars throughout the country being quickly converted. In fact, before the inspectors had delivered their judgement, Accrington had stipulated that Rawtenstall cars using its tracks were to be adapted, being fearful of a runaway on the steep Manchester Road.

In 1912, two more double-deckers arrived, accompanied by six single-deckers, for the Water line, being fitted with magnetic and slipper brakes for this steeply graded section. No more additions were made until 1925, when eight Brush-built double-deckers brought the fleet strength up to 32. Two years later, the first motorbuses appeared on the town's streets, and in 1929 the decision was taken to abandon the tramway, a process which, as should now be clear, had, like the inception, to be a joint one. The last car ran on the Bacup, Loveclough and Water sections on 31 March 1932, the service to Accrington having gone two years before.

Before winding up this section, reference must be made to the interlopers mentioned at the beginning of the chapter. Rochdale Corporation reached Bacup with its electrified line via Whitworth, but terminated in Bridge Street, a tantalising few yards away from the Rawtenstall line in Market Street. Ramsbottom UDC (swallowed up into Greater Manchester in 1974 and a SELNEC constituent in 1969) reached Edenfield (now part of the Borough of Rossendale) in 1913, not by tramcar as originally envisaged, but with trolleybuses. They lasted until 1931.

Below: Could this be the tramways manager posed alongside Rawtenstall's UEC car No 14, now vestibuled? All these saw their Raworth regenerative controllers replaced by the normal Westinghouse type following an accident in 1910, on the Loveclough section, when the Board of Trade cast doubts on the supposedly effective use of electricity, accredited to them. *Roy Marshall Collection*

Above: A broadside view of either No 9 or No 12, both of which were fully enclosed in 1926. *National Tramway Museum*

Above right: The impending encroachment by the motorbus is illustrated here on a later Rawtenstall ticket. *National Tramway Museum*

Right: Along with two balcony cars, six single-deckers were purchased by Rawtenstall in 1912, again built by UEC. They were intended for the steeply graded line to Water, and were equipped with magnetic and slipper brakes. This is No 23, photographed at Preston prior to delivery, complete with curtains. *National Tramway Museum*

Below right: In preparation for tramway abandonment, Rawtenstall purchased 30 Leyland motorbuses between 1930 and 1932. After final closure in 1932, a ceremonial farewell was afforded the cars on 7 April, and here a TD2 Titan poses with car No 26, one of the final eight built by Brush in 1921. The twin trolley poles were believed to be unique in Lancashire. Because of their size, these 72-seat cars were restricted to the Rawtenstall-Bacup section. Completing the line-up is former Rossendale Valley Tramways Co Thomas Green steam tram engine No 6, one of two retained by the corporation as snowploughs. This one survived until World War 2, when it sadly succumbed to the frantic search for scrap metal. *Roy Marshall Collection*

100

CHAPTER 7

END OF THE CHAIN

On the outskirts of Colne, on the road to Keighley, a large building used as a commercial vehicle repair shop belies its antecedents as a tram shed. Anyone exploring behind it would come across a sharply descending, twisting cobbled footpath, at the bottom of which is a filling station, 'Tram Track Motors'. Here, and a mile further on, where a similar footpath is encountered, are the remaining pieces of evidence of the one-time grandly named Colne & Trawden Light Railway.

Colne, neighbouring Nelson and the larger town of Burnley, were for many years constituent members of the northeast Lancashire amalgamated transport undertaking — the Burnley, Colne & Nelson Joint Transport Committee.

From Burnley, the outer suburb of Harle Syke is reached by way of Briercliffe Road, where, not far from the town's hospital, is a public house, ordinary in many respects and perhaps 'run of the mill' as far as the average imbiber is concerned. Except to some, as it is named after the Russian warships during the Russo-Japanese war, when they attacked British fishing boats: the 'Baltic Fleet'. Or is it? The sign swinging over its door depicts a steam tram, one of Burnley's 'Baltic Fleet', belching smoke from its long smoke stack in a menacing fashion, like a Russian warship. These are all relics and reminders of the area's transport heritage.

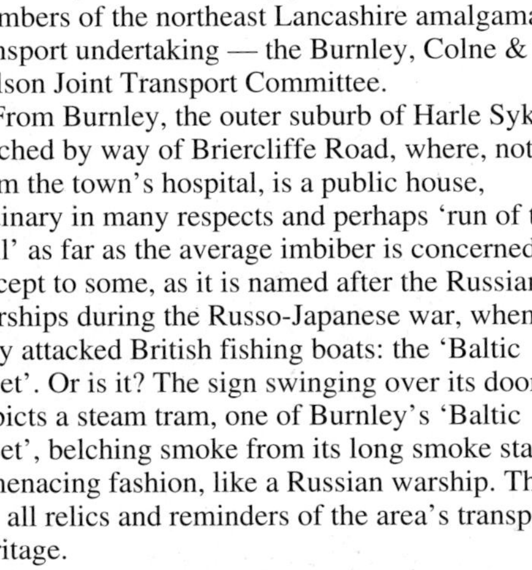

Midway along the valley of the River Calder between Colne and Burnley, lies Nelson. Although similar, this cotton town is, however, the baby of the three. It had sprung up as a child of King Cotton from a cluster of hamlets and derived its name from the hostelry, the Nelson Inn, which stood at the junction of the turnpikes to Burnley, Barrowford and Colne. All three towns possessed their own transport undertakings prior to the 1933 Joint Committee formation. This exists at the time of writing, albeit as a council-owned bus company, Burnley & Pendle Transport Ltd, Nelson and Colne having been brought together in 1974 as parts of the borough named after the hill, famous for its witches, which overshadows the area. Trams survived just into joint ownership to operate as BCN cars, all with the distinct characteristics of their individual pedigrees, which can be examined in turn.

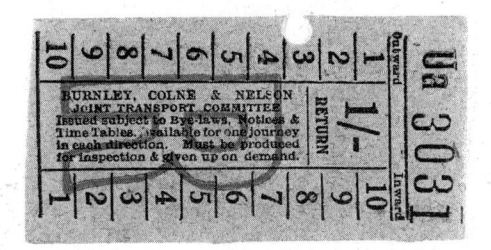

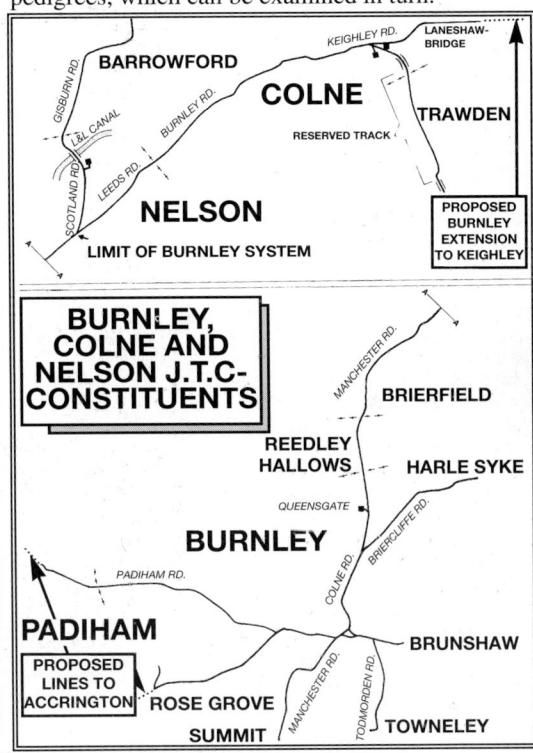

Above: A BCN ticket. *National Tramway Museum*

BURNLEY

A London concern, the Tramway & General Works Co Ltd, was responsible for establishing the first tramway in the area, based in Burnley. Steam-hauled, and to the 4ft 8½in gauge, it opened on 17 September 1881 to replace the Burnley-Nelson section of a horse omnibus service, which had plied as far as Colne since 1861. Within Burnley, there had also been a route from Wood Top down to Barden Lane. The tramway was also projected to run on the other side of Burnley, to Padiham, and on the commencement of operations, a new company, the Burnley & District Tramways Co Ltd, was formed.

As was the case in Blackpool, it would be correct to say that this area never saw a horse tramway, but it would be equally in order to observe that it did have horse-drawn tramcars, because the company's steam tram locomotives regularly failed. In fact, this even happened during the first day of trials, when horses had to be summoned to bring in an ailing machine from Westgate, back to the Queensgate Depot. This unsatisfactory state of affairs was exacerbated by the apparent unease of the Board of Trade, who initially refused permission for the trams to run along Church Street, stating they would be dangerous in such a narrow and crowded thoroughfare. After the company agreed to a four miles per hour speed limit in Church Street and to eight miles per hour elsewhere, the inspector relented. The 'Baltic Fleet' soon incurred the wrath of Burnley Council, though, who were far from pleased with the smoke and noise from the five Kitson locomotives, and, once again, in April 1882, horse power took over. To try and overcome the problem, a new Falcon locomotive built in Loughborough was purchased, but this disgraced itself, especially in the eyes of the locals in the Tim Bobbin Inn (named after a local poet) on Padiham Road, when its boiler exploded outside on its first run. Nevertheless, 11 more Falcons were bought, but it was to be another three years, in 1885, before problems were satisfactorily ironed out and the horses could be retired.

The corporation had by that time widened some streets but, until that work was complete, the steam trams had been confined to the Queensgate to

Below: Burnley District Tramways Co's early efforts at providing an acceptable service proved something of a disaster, with its initial five Kitson tram engines failing to come up to scratch. Here, No 4 with open-top trailer car traverses Colne Road. Dr Lovelace's soap can no longer be purchased at the corner shop (nor anywhere else, for that matter) but Nestle's products and the Burnley Express are still to be had in what is, possibly, now an 'open 'til late' mini-market.
Roy Marshall Collection

Above: Twelve more successful Falcon locomotives were taken into stock between 1883 and 1897. This is No 13 of 1885. The trailer was an eight-wheeler (No 6) built by Starbuck of Birkenhead in the same year. Already, the nickname 'Baltic Fleet', like in other East Lancashire towns, was catching on, as is explained in the text. *National Tramway Museum*

Below: Built in 1901 for the electrification of the system, Milnes car No 16 was, however, photographed post-1914, as its 'Burnley Bogies' acquired in that year can be seen. Interestingly, its original Brill bogies were rebuilt as Burnley types, like the rest of its 37 sisters. It was later to be modified by the fitting of a top cover. *Roy Marshall Collection*

Above & below: A comparison between the more usual 'equal wheel' bogie and Henry Mozley's 'Burnley Bogie' is useful at this point. The small pony wheel was designed to enable safer cornering at high speeds. Both UEC and Mountain & Gibson of Bury manufactured them under licence. *Roy Marshall Collection*

Nelson section, horses filling the breach through the town centre. It was not until March 1883 that steam running between Nelson and Padiham was possible.

The seven-mile journey took 20min (40min by bus now!) and eventually 12 pairs of locomotive and trailer, plus one spare locomotive, were taken into stock, later to be augmented by further deliveries. (It is likely the five Kitsons were dispensed with before 1883.) Unusually (and uncomfortably) some trailers were open-top, but passengers could take cold comfort from the fact that reduced fares applied when riding up there!

Under the provisions of the 1870 Tramways Act, Burnley Corporation could take over the system in 1900, but, completely fed up with the often unreliable and dirty steamers, it obtained Parliamentary powers in 1898 to pre-empt matters, and bought it for £53,000. The Corporation intended to electrify and reduce the gauge to 4ft in common with its neighbours. (There had been plans to link up later with Accrington, under the auspices of the Blackburn, Padiham and Whalley, and the Accrington and Burnley Light Railways, both authorised in 1901. These came to nothing and the three towns were to keep themselves to themselves; another lost opportunity was the adventurous proposal to meet up with the Keighley system in the days before the Nelson and Colne lines were conceived.)

In 1901, Nelson Corporation, the Urban Districts of Padiham and Brierfield and, incredibly, on reflection, Reedley Hallows Parish Council (a tiny strip between Burnley and Brierfield) all notified their intentions to exercise Tramways Act options, and, as a result, acquired the sections owned by Burnley within their townships. (This made Reedley Hallows the only tramway-owning parish council in the country.) It was their intention for Burnley to carry on with its modernisation of the tramway on their behalf, and, likewise, operate it. Most of the old system had been single track, and work got under way in the spring of 1901 in doubling it, erecting overhead and re-laying the track to its new gauge.

The Padiham side's conversion was completed in the latter half of 1901, and, with the opening of the Nelson route early the following year, through running between there and Padiham was introduced. This was to be the corporation's 'main line', a name perpetuated to this day in bus operation.

Below: Top-deck windshields were fitted to the 14 Milnes cars received in 1903. They were otherwise generally similar to the first batch. Once again, the 'Burnley bogies' date the picture as post-1914, but before 1920, when the last of them became top-covered. This is No 32. *Roy Marshall Collection*

Right: Manchester Road, at the town centre end of the line up to the Summit is the location of this 1930s roof-top shot of two single-deckers about to pass. Nearest the camera is No 69, one of five cars purchased in 1921 from English Electric. They were mounted on the now inevitable 'Burnley bogies', some of which were manufactured by the corporation themselves. *Roy Marshall Collection*

Above: Eight single-deckers were also bought in 1903, from ER&TCW, seating 44. No 40 passes the Yorkshire Hotel *en route* to Towneley. This line opened in 1904 as Burnley began to expand its system away from the Padiham-Nelson 'main line'. *Neville Lockwood Collection*

Right: The five Hurst Nelson cars delivered in 1909 were of the balcony variety, as depicted here by No 50 in the yard at Queensgate depot. They seated 71. *Neville Lockwood Collection*

Left: One of the 1909 balcony cars climbs away from Lanehead towards the Harle Syke terminus. The former point was the original extent of the line reached in 1910 which left the main line at Duke Bar on Colne Road, and it was to be another two years before the eventual terminus was reached. This involved the construction of a new road, on to which the car is entering, to alleviate the sharp right hand bend visible in the middle distance. The new road was the only part of the route double tracked, the start of it being where the photographer was standing. *Neville Lockwood Collection*

Twenty-four double deck, open-top cars, seating 64, were bought for the inaugural route. Manufactured by Milnes, they wore a maroon and cream livery, contrasting with the drab black and white of the steamers. Once the conversion was complete, work was started on tramways within the borough boundary. The expanded network soon opened its new routes: to Rose Grove in July 1903, and to Towneley Park and Summit (Manchester Road) in February 1904. In 1910, these latter two were extended to Rock Lane and Rossendale Road respectively. Also in 1910, two more new routes were inaugurated. In October, Gannow Lane got its service, followed by Lane Head in December, which reached Harle Syke on 13 August 1912. Most of these town services were maintained by rather plush single-deckers, 13 being introduced between 1903 and 1912.

Double-deck strength was increased during this period by the addition of another 23, giving a fleet total of 67. This, of course, was all at the zenith of Harry Mozley's reign at Queensgate, as manager of the Corporation Tramways and the locally acclaimed 'Tramway King', an accolade no doubt stemming from his prominence nationally on the tramway scene, being at one time the President of the Municipal Tramways & Transport Association, and a Fellow of the Royal Society. He had managed to squeeze a degree of reliability out of the recalcitrant steamers in his earlier days, and had been present at their birth; by 1882 he was company secretary. Mozley was to stay in charge until 1930, and amongst his many ideas was the invention of the 'Burnley Bogie', which enabled speedier and safer cornering. The idea was patented, and manufactured by UEC under licence. Most of the Burnley cars ended their days with them, either being fitted from new, as were those from 1910 onwards, or rebuilt with them. He also devised the right-hand spiral staircase, which broke away from the more usual one rising to the left. One of the advantages of this design was the increased room on the platform. Another five cars, all single-deck, were added in 1920, bringing the fleet strength to 72.

A late extension was a new line to Stroyen Street at Brunshaw in 1927, by which time Parliamentary powers had been obtained (in 1921) to operate buses beyond the boundary, not exceeding three miles from the Town Hall. The first service was initially experimental, between the Cattle Market and Stoneyholme, later extended to Reedley Halt on the Skipton railway line. Leyland A13s with Leyland's own bodywork were bought at first. The five of them were only 6ft 6in wide, with a seating layout resembling the trams', and had tram-like rear platforms for speedy loading and unloading. The lure of the omnibus, becoming so familiar all over the land, was now taking hold in Burnley.

Total tramway mileage was 13.5, which included the 2.7 miles the corporation leased, but on 4 April 1932 the first abandonment came about, with the last cars on the through Harle Syke to Rose Grove service, at a time when the mechanism was already in place for the now renamed Burnley Corporation Tramways & Omnibus Department to become the major constituent of the joint committee. It was to transfer all but four of its cars to the new ownership, the 1932 conversion having reduced the requirement.

BURNLEY CORPORATION TRAMWAYS.

SEE MAP ON OTHER SIDE—shewing connections with Light Railways and Omnibus undertakings.

Time Table of First & Last Cars.

From BURNLEY CENTRE TO	SUNDAY a.m. / p.m.	MONDAY TO FRIDAY a.m. / p.m.	SATURDAY a.m. / p.m.	
NELSON	8 50 / 10 30	5 50 / 11 0	5 50 / 11 30	
PADIHAM	8 35 / 10 30	5 0 and 5 35 / 11 0	5 0 and 5 35 / 11 30	
HARLESYKE	9 5 / 10 35	5 55 / 11 5	5 55 / 11 35	
ROSEGROVE	8 45 / 10 35	5 15 / 11 5	5 15 / 11 35	
TOWNELEY	8 45 / 10 35	5 50 / 11 5	5 50 / 11 30	
MANCHESTER RD. SUMMIT	8 50 / 10 40	5 50 / 11 10	5 50 / 11 30	
PARK LANE	— / 10 35	— / 11 5	— / 11 35	
From (To BURNLEY CENTRE) NELSON	9 0 / 10 0 (a) 10 50	6 0 / 10 30 (a) 11 0	6 0 / 11 0 (a) 11½ 30	(a) thro' to Pdihm
PADIHAM	9 5 / 10 5 (e) 10 55	5 25 and 6 0 / 10 35 (e) 11 25	5 25 and 6 0 / 11 5 (e) 11 55	(e) thro' to Nelson
HARLESYKE	9 5 / 10 55	5 45, 6 0 and 6 15 / 11 5	5 45, 6 0 and 6 15 / 11 25 (b)	(b) to Depot 11–55
ROSEGROVE	9 5 / 10 55	5 35, 45, 55, and 6 15 / 11 25	5 35, 45, 55, and 6 15 / 11 55	
TOWNELEY	9 0 / 10 50	6 0 / 11 20	6 0 / 11 45	
MANCHESTER RD. SUMMIT	9 5 / 10 55	6 0 and 6 15 / 11 25	6 0 and 6 15 / 11 45	
DEPOT & BURNLEY CENTRE	(c) 8 25 / (d) 11 20	(c) 4 50 / (d) 11 50	(c) 4 50 / (d) 12 20 a.m.	(c) to B'ly (d) Depot

LAST CARS in connection with Burnley Cars in Nelson Centre.

FROM LANESHAW BRIDGE	— / 9 55	— / 10 15	— / 10 32
FROM TRAWDEN	— / 9 45	— / 10 5	— / 10 36
FROM BARROWFORD	— / 10 10	— / 10 45	— / 11 10

Every endeavour will be made to run Cars as shown in this Time Table, but the Corporation of Burnley do not guarantee the service. They reserve the right to alter the running of Cars without notice, and will not be responsible for any loss or inconvenience which may be caused by the Cars not being run in accordance with this Time Table. The services of the Cars are from ten to two-and-a-half minutes according to the requirements of the traffic. Special Cars are run for the workpeople as required, and any complaint of inadequate service will receive immediate attention. Through Excursion Cars are run as required between Burnley, Barrowford, Trawden & Laneshaw Bridge. Omnibuses run between Padiham and Whalley, Summit and Loveclough, Burnley and Worsthorne, and other districts, but the Corporation have no responsibility for these services.

Parcels are carried cheaper and quicker by the Cars than by any other means.

General Office : Queensgate, Burnley, Nov. 3rd, 1921.

Left: A Burnley Corporation Timetable of First and Last Cars. *National Tramway Museum*

Above: A second maintenance vehicle was built by the corporation in 1925 to augment its first (also built in-house in 1903). Their numbers, 1 and 2, clashed with the car fleet, but they were officially classed as locomotives, thus warranting a separate series. *Alan Catlow Collection*

Above: Milnes car No 10 of 1901 is seen here in Burnley centre in the early 1930s, after having been renumbered 68 in 1926, following the misfortune of having two accidents at Lanehead, and probably to confuse nervous passengers. The single-deck car with that number became No 73. The Leyland PLSC1 Lion with rare Knape body is No 12, about to depart for Rose Grove on the tramway avoiding route via Sandygate.
Alan Catlow Collection

COLNE

Colne, like Nelson, was a latecomer to tramway operation. Under the provisions of the Light Railway Act, three separate schemes were put forward. Colne Corporation and the Leeds firm of Batley and Greenwood's proposals were for a line from Nelson to Trawden, a cotton weaving village of reasonable size, to the southeast of the town, whilst Nelson Corporation's was for one from that town to Colne. All would be electric. (As will be seen soon, Nelson was laying down plans for its own tramways and this proposal featured in them.)

However, Colne Corporation collaborated with Batley and Greenwood, who were to build, operate and, indeed, own the line, but the Council would have an option to purchase later. Subsequently, the Colne & Trawden Light Railway was registered in 1901, and the Parliamentary powers obtained also included approval for a branch line, leaving the main line at Heifer Lane, to Laneshawbridge on the road to Keighley, another link of course in the earlier proposed Burnley chain. The legislation did not, though, permit the laying of tracks into the Borough of Nelson and therefore the railway's line had to finish at the boundary near Bott House Lane. In total, 5.23 miles of track were to be laid.

In December 1905, the lines from Bott House Lane, through Colne to Lane House Lane in Trawden and to Emmott Lane, Laneshawbridge were completed, having been opened in stages: Heifer Lane (where the depot was situated) to Queen Street on 28 November 1903; to the Nelson boundary on 30 November; to the Rock Hotel, Trawden by June 1904; to Laneshawbridge by the December; and finally to the Zion Chapel on Lane House Lane. Despite its railway title, all its tracks were conventional street tramway, with the exception of the reserved sections mentioned earlier, known locally as the 'Snake Track', and the more straightly aligned and obviously named 'Tram Track'. This latter section ran from the Rock Hotel to the terminus, both thus avoiding the notoriously treacherous hills on Cotton Tree Lane and Church Street.

The company required 12 cars to operate its line, six arriving in 1903, the last, three years later. All were 50-seat open-toppers of Milnes, Brush or

Below: The first section of the Colne and Trawden Light Railway, between Heifer Lane and Queen Street, was opened on 28 November 1903. One of the first six Milnes 50-seaters descends Albert Road towards Primet Bridge and the railway station. *Neville Lockwood Collection*

Above: The first tram to Trawden! Church Street, near to the Rock Hotel, was the initial point reached before the last section of line was laid to the Zion Chapel in Lanehouse Lane in December 1905. The car attracts interested village folk on its arrival from Colne. *Neville Lockwood Collection*

Below: To reach the Zion Chapel, a reserved track was constructed, seen here. Like the one at the Heifer Lane end of the Trawden section, it is discernible at the time of writing and is in use as a footpath. *Neville Lockwood Collection*

Milnes Voss manufacture. Although Trawden and Laneshawbridge hosted the occasional arrival of Burnley cars — working excursions for its townsfolk to sample the rural pleasures around Wycoller and the fringes of the Brontë country — Colne & Trawden's passengers were obliged to change to Nelson Corporation cars at the boundary, a situation which persisted until 1911, when through working between the two towns was agreed upon. However, rebooking at the boundary was insisted upon for many years after that.

The Colne & Trawden directors saw a good return on their shareholders' investments, and at one time considered increasing the fleet size, but decided instead to canopy half of it, thus making travel more comfortable and attractive. Conflicting records show the work having been done in 1911 by either Brush or Milnes Voss.

Colne Corporation purchased the system in its entirety, despite part of it being in the Urban District of Trawden, on 24 March 1914 — not, as history shows, and has been demonstrated in the preceding chapters, the best of times to buy a tramway. Its title was changed to Colne Corporation Light Railways and T. G. Richardson came from Bradford as the new manager. He immediately took stock of the new acquisition for the corporation and recommended that one new car was needed — a UEC-built double-decker delivered in December. Subsequently, he also brought about the fitting of canopies to the remainder of the open-top cars. (Although a beautiful part of the region, its winters can be harsh; 'one top-coat colder' being the comparison sometimes made with more far westerly parts. The now enclosed cars would have been most welcome!)

Needless to say, World War 1 left the fleet run-down and in desperate need of replacement, or at least complete renovation, a programme of which was got under way in 1919. Eventually, five new cars were purchased, the first pair being delivered to Heifer Lane in 1921. They were equipped with 'Burnley Bogies', manufactured by English Electric, the cars' builders. The remainder, the first totally enclosed cars, came from Brush in 1926.

The 1920s cotton slump witnessed the system's meagre profits being hived off to support the corporation, which was short of cash. The tram service, despite its not too certain financial future, had to provide 10% of income to the Borough Treasurer. Times were not good. Bus competition was looming, the Colne & Earby Omnibus Co was

Below: The rural branch line to the village of Laneshawbridge on the Keighley Road was completed from Heifer Lane in 1905, the first tram running simultaneously with that over the Trawden extension. To permit this, four more cars were added to the fleet, similar in design to the first, but built by Brush. Here is No 9 at the far terminus, Emmott Lane. By this time, through working via Colne over Nelson's system was possible. *Alan Catlow Collection*

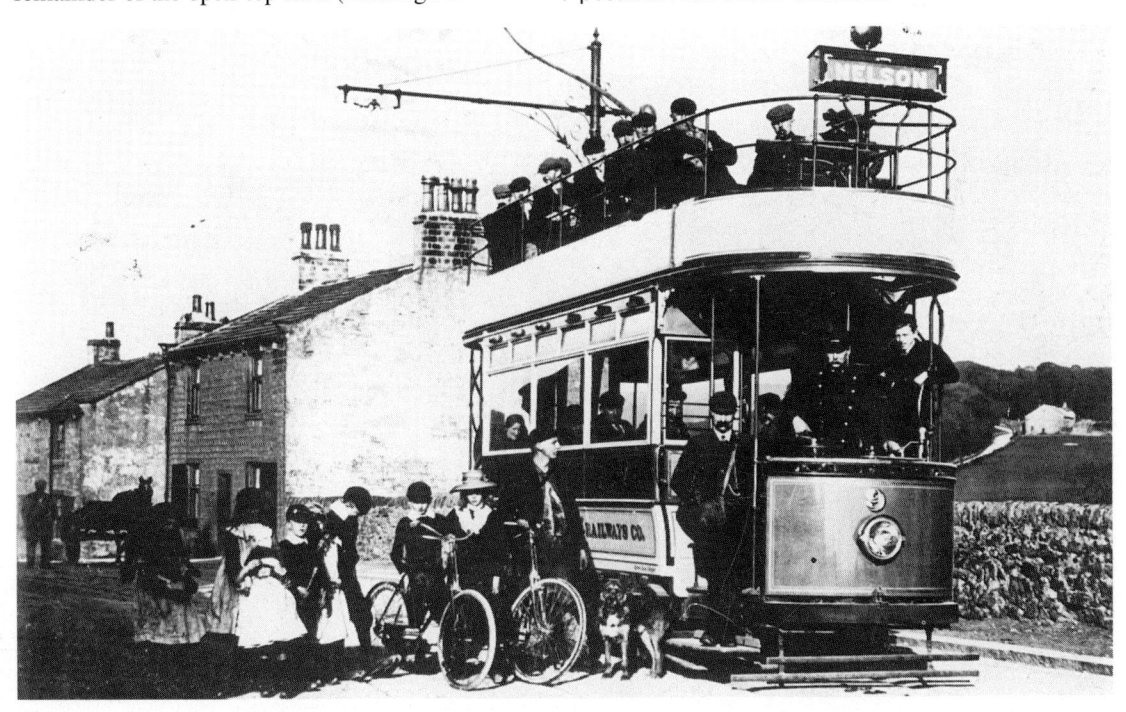

111

Above: Another Laneshawbridge scene, with one of the pair of 1906 Milnes Voss cars exchanging passengers with a local horse bus and Ezra Laycock's double-decker bound for Cowling. Laycock is reputed to have been the first motorbus operator in Yorkshire, who later sold out to BCN, only to quickly be re-established on a service between Barnoldswick and Skipton. The firm sold out to Pennine of Gargrave in 1972. It is difficult to date this photograph. The bus appears to be a Commer, WR 1622, new in 1919, but the fashions look more early 1900s. Laycock had a Milnes-Daimler double-decker in 1906, but it bore no resemblance to this. *Roy Marshall Collection*

Below: Primet Hill, Colne, in 1913, by the joint Lancashire & Yorkshire/Midland railways station, its bridge dictating the use of low-height cars by Colne and Nelson. A Colne & Trawden car heads through the rain to Laneshawbridge. Here, and in Market Street, were the only double track sections beyond Nelson centre. *Neville Lockwood Collection*

but one protagonist in the arena with others impatiently waiting in the wings. That said, Colne Corporation was itself to become an enthusiastic bus operator. In 1920, three second-hand Tilling Stevens chassis, fitted with new bodies, had been bought. They were pressed into service as occasional tramway replacements, in the event of power failures, but permission to operate a regular service was not forthcoming and they were soon sold. Shortly after, in January 1923, the first of four routes was established and by the end of the following year seven buses were in stock, more than in either the Burnley or Nelson fleets at the time. This had rendered depot space inadequate and in 1921 a new shed had been built in Stanroyd Road, adjacent to the original, which remained in use until demolished a few years before Burnley & Pendle's Colne depot's final demise in 1981. In 1923 or thereabouts the cars' royal blue and white livery was superseded by maroon and cream.

In response to outside bus pressure, and to economise in those lean years of the 1920s, two cars were converted for one-man operation by having their upper decks and stairs removed. This occurred in 1924 and 1925, the scheme not particularly impressing the Board of Trade Inspectors, who doubted its effectiveness.

Laneshawbridge lost its trams on 19 October 1926, which came as no surprise, as the branch served a sparsely populated area. It was at this time that the corporation was attempting to purchase the bus operations of Ezra Laycock of Cowling, who ran from Keighley as far as Laneshawbridge. With the take-over successful, from April 1927, Colne Corporation was able to work the long through service to Keighley as an extension to its replacement for the Laneshawbridge cars. On 3 June 1928, the Heifer Lane-Trawden route was abandoned, although peak workings, as far as Cotton Tree, continued down the 'Snake Walk'. All that now remained was the main line from the tram sheds towards Nelson, but in 1932, by which time amalgamation was in the air, an express bus service between Colne and Padiham had started to drive the final nails into the Colne trams' coffin. The tramcar could not fight back, as the section between Colne and Nelson was single track, which would have created scheduling problems for a more intensive retaliatory service, and the track through the town centre between Heifer Lane and the Queen's Hotel at the foot of Primet Hill was becoming urgently in need of replacement. It was nearly time to call it a day, but not just for the time being, though.

Below: Balcony car No 13 was a 'one-off' in 1914, and the first bought by the line's new owners, Colne Corporation. Built by UEC, it seated 52 and displayed the coat of arms of the borough, which still chose to use 'light railways' in the title. *Neville Lockwood Collection*

NELSON

Pomp and civic ceremony heralded the opening of the Nelson Corporation Tramways on 23 February 1903, nine months ahead of Colne's first section. Bedecked in garlands, with the mayor, Dr W. Jackson, at the controls, one of the first six Brush open-top 40-seat double-deckers, delivered the previous year, crept out of Scotland Road towards the Colne boundary along Leeds Road. (The borough's mayor travelled on the last car in 1934, without pomp and garlands, so representative of the falling from grace suffered by the tramcar in so many towns, where not long before it had been their pride and joy.)

Of course, Burnley cars had been working over Nelson-owned metals into the town for some months, but now Nelson had some of its own. As far back as 1882, powers had been secured to build a steam tramway from the George & Dragon in Barrowford, along Gisburn Road, up Reedyford Brow and into the town centre, to meet the Burnley steamers, no doubt with the prospect of through running. The scheme joined the throng of those never to materialise, and it was not until some twenty years later that construction of a corporation tramway began in earnest. Approval for two routes had been granted under the Light Railways Act: from the town centre, along Leeds Road to the Colne Boundary; and to Higherford Bridge at the Blacko end of Barrowford. Barrowford UDC, itself young and go-ahead, had also obtained necessary powers, so it can be reasonably assumed that this route was to have been theirs, which maybe would have added Barrowford to what became the small select band of English Urban District bus undertakings — West Bridgford, Cleethorpes and Ramsbottom. In the event, Nelson was to build and operate the line through the village.

A depot was built in Charles Street on the bank of the Leeds & Liverpool Canal and adjacent to the Nelson Electricity Works, from where power was obtained for the system. It had been intended, at the time of drafting the legislation, to have the car sheds near to Hallam Road, but this would have involved the laying of a spur of a mile or thereabouts off the Colne line on Leeds Road, so the option of erecting the building on the corporation's own land at Charles Street was taken up. The original 1903 structure was, however, later demolished and a new, larger shed built, which was still in council use in 1994 as Pendle's engineering depot.

The six double-deckers, found to be hard pressed, with no fewer than 674,786 passengers being

Right: Nelson Corporation Tramways No 1 waits in the loop in Scotland Road, bound for Barrowford. This was one of six Brush open-top cars acquired in 1902 and 1903 for the opening of the system in the latter year.
Roy Marshall Collection

carried in the first six months of the system, were joined by a trio of ER&TCW single-deck combination cars, which had open smokers' compartments at either end, in late 1903 and 1904, all wearing a red and white livery, later changed to brown and cream. These were now considered sufficient to maintain services on the 2.75 miles of line. (This did not include the 0.76 mile leased to Burnley.) All of it was single track with passing loops, except for short double sections between the Higherford Bridge terminus and Bankhouse Street, and between Bank Street and the Fleece in Barrowford, and on Reedyford Brow (later extended in 1906 to Charles Street) in Nelson.

Two additional cars, UEC balcony types, were purchased in 1912 to meet increasing demands stemming from the agreement reached the year before with the Colne & Trawden company for through running. This pair were of 'lowbridge' design, to enable them to negotiate the railway bridge by Colne station. Six more to this configuration, again built by UEC, were bought in 1916 to replace the initial cars, the last three of which were also being finally usurped by new arrivals, this time by the corporation reverting to Brush for 'lowbridge' balcony cars in 1925.

A couple of years before, the corporation had wisely exploited the need to provide people living away from the tramway with access to it, and by September 1923, after the necessary powers had been secured, its first 'feeder' bus routes had been opened. Three commenced initially and were all one-man operated, until 1932, when the 1930 Road Traffic Act prohibited the practice. These bus services, however, remained as feeders until the end of the tramway, which soldiered on intact until the amalgamation.

Right: Sister car, No 2, at Reedyford on the Nelson/Barrowford boundary. A feature of Nelson cars was the decency panel protecting female passengers climbing the stairs.
Neville Lockwood Collection

114

115

Top: Ninety years after this photograph was taken, a seat at the junction of Gisburn Road and Colne Road, Barrowford, outside the old toll house, is still there. Certainly not the same one, it however features prominently in a scene that is much the same today, save for the increase in traffic and the long gone Nelson Corporation Tramways No 4, heading towards its home town. The building on the left is the George & Dragon, whilst the terminus at Higherford is a hundred yards or so round the bend. Although in use here as a shop, the toll house (originally built to service the Nelson to Gisburn turnpike) later became a dwelling and is now an annexe to Pendle Heritage Centre, just over the bridge. *Neville Lockwood Collection*

Above: An ER&TCW 38-seat combination car was added to the Nelson fleet in 1904 and numbered 9. At this time, the livery was red and white. Like neighbouring Colne, Nelson later adopted a more mundane colour scheme, brown and cream. *National Tramway Museum*

Left: The second Nelson car to carry the fleet number 1, was one of six 'lowbridge' cars constructed by UEC in 1916. The small 'N' suffix in this Albert Road, Colne, view dates it as after the joint committee formation. *National Tramway Museum*

Right: UEC supplied a water car/sweeper in 1905, and it was given the fleet number 12. The former Burnley car just visible confirms this to be an early post-amalgamation view of the Nelson depot in Charles Street. It still stands, as part of Pendle Borough Council's engineering department. *Roy Marshall Collection*

Left: Another Nelson tram in Colne on the through service. Second Generation No 7, a Brush 55-seat balcony car of 1925 (low-height, of course, to negotiate Primet Bridge), on the passing loop by the Co-op in Albert Road, again, in earlier BCN days. Cobbles, trams and tram lines excepted, this view has changed little over the years. *Roy Marshall Collection*

BURNLEY, COLNE AND NELSON JOINT TRANSPORT

A four line paragraph in the minutes of the meeting of the Lancashire & Cheshire Tramways Authorities Council on 19 May 1933, held at the Victoria Hotel in Manchester, read 'The Secretary reported that consequent upon the formation of a Joint Board, the Colne and Nelson Tramways had withdrawn from membership of this Council'. Joint Boards had been discussed at the previous meeting in October 1932, when the resolutions of the Municipal Tramways & Transport Association's Conference at Eastbourne, the month before, were reported upon. Basically (and radically) it had been agreed that municipal authorities 'should combine to form Transport Boards within certain defined areas, such areas to be decided with due regard to geographical situation'. Area Councils were asked to consider the proposals and the members of the Lancashire & Cheshire Council were unanimous in their support.

Burnley, Colne and Nelson had pre-empted these ideas by some two years, as in May 1930, the Town Clerk at Burnley had written to his opposite numbers, suggesting they meet to discuss a scheme to pool their resources. It was met with enthusiasm, and the wheels were set in motion towards the formation of a Joint Board or Committee. At this time Colne Corporation was promoting a bill, dealing with other matters, through Parliament, and it was thought prudent to include the transport amalgamation within it. Thus, the Colne Corporation Act, 1933, authorised the establishment of the Joint Committee, from 1 April 1933, with its headquarters at Queensgate in Burnley, the site of the first steam depot, gradually developed over the years. Representation on the committee was proportional; four councillors from Burnley and two each from Nelson and Colne, as was the responsibility for losses and the sharing of profits; Burnley 19/32nds, Nelson 8/32nds and Colne 5/32nds.

Below: Ex-Burnley Corporation No 65 was one of the many never to receive the new scroll, and is seen here, still adorned 'Burnley Corporation Tramways' alongside other withdrawn cars in the yard at Queensgate in 1935, the year of abandonment. The height difference with the two Nelson cars can be seen. *Alan Catlow Collection*

Apart from the Queensgate Depot, the Committee, soon to be widely known simply as 'BCN' (or by staff, quaintly, as 'The BC and N'), inherited the others, Colne's at Heifer Lane and Nelson's in Charles Street. Eighty-eight trams are believed to have come into joint ownership, 68 from Burnley and 10 each from the other two, along with 47 buses from Burnley, 24 from Colne and 15 from Nelson, all single-deck, except for 16 from Burnley. With immediate effect, suffixes were added to fleet numbers: Burnley — B; Colne — C; Nelson — N.

Tram routes were Padiham-Nelson via Burnley Centre, Nelson-Colne, Nelson-Higherford, and, in Burnley, Manchester Road-Townley and Manchester Road-Brunshaw. There were two additional extra relief workings: Brierfield-Nelson and, in Burnley, Park Lane-Brennand Street. Bus routes had grown rapidly and 18 were running on formation. One, interestingly, was the express service referred to in the Colne section, Burnley's Padiham-Colne, duplicating the tram services of all three authorities.

Now duly formed, the BCN committee pressed on with its expected modernisation programme, the cornerstone of the whole initiative. 'Corporate image', to use modern jargon, saw the adoption of a new livery, crimson lake and primrose. The burning question, though, was what to do with the trams. A review of the system revealed the former Nelson and Colne lines to be in the worst state of repair, whereas Burnley had renewed much of its in the 1920s. The absence of any respectable lengths of double track in Nelson and in Colne did not help their cause, and the low bridge by Colne station precluded all but single-deckers from the former Burnley fleet from through working, which restricted the use of four Burnley double-deckers transferred to Nelson for the Higherford service. The Committee felt that the Burnley routes could struggle on for a while, but decided that buses would replace the cars on the Colne and Nelson sections from 7 January 1934. Charles Stafford, the Manager, being concerned that tram drivers would find it difficult in coming to terms with having to learn to drive vehicles which would go where they wanted, tried to make the transition easier by ordering Leyland buses (after the initial 18) with torque convertors (or 'self changing gears'). On 14 October, the Burnley local routes were abandoned, leaving only the original 1881 route from Padiham to Nelson running. In 1931, Stafford had drawn up plans to convert it to trolleybuses, but it was to be 32 Leyland TD4cs that replaced the cars on 7 May 1935, bringing tramcar operation in this northeast corner of Lancashire to an end, and although cars elsewhere rolled and clattered on, it also brings to a close this story of the tramcar in the northwest, hopefully, 'thus far'!

Above: The formation of the Burnley, Colne & Nelson Joint Transport Committee on 1 April 1933 saw the introduction of a new scroll, incorporating the three constituent authorities. *Neville Lockwood Collection*

Left: The new BCN scroll can be seen on the side of the former Burnley Corporation car No 41, at the Summit. Basically, the old Burnley livery was perpetuated by the new joint committee formed in April 1933, although no Nelson or Colne cars were ever repainted; the writing was already on the wall. *Roy Marshall Collection*

FLEET LISTS

ACCRINGTON CORPORATION TRAMWAYS — FLEET LIST

Date	Fleet Nos	Type	Builder	Seats	Truck(s)	Motors	Controllers	Notes
1907	1-4	Single-deck	Brush	32	Brush	Brush	Brush	
1907	5-18	Balcony	"	22/28	"	"	"	a
1908	5,6	Single-deck	"	32	"	"	"	
1909	21, 22	Balcony	"	22/28	"	"	"	
1909	23	Single-deck	"	32	"	"	"	
1910	24, 25	Balcony	"	22/28	"	"	"	
1912	26	"	"	"	"	"	"	
1912	27	Single-deck	"	32	"	"	"	
1915	28-30	Enclosed	"	32/44	Brush bogies	Dick, Kerr	Dick, Kerr	
1920	31,32	"	"	"	"	EE	EE	
1926	42,43	Enclosed 'Lowbridge'	"	36/34	Peckham	"	"	

Notes: **a)** 5 and 6 renumbered 19 and 20 in 1908

RAWTENSTALL CORPORATION TRAMWAYS — FLEET LIST

Date	Fleet Nos	Type	Builder	Seats	Truck(s)	Motors	Controllers	Notes
1909	1-16	Balcony	UEC	22/29	Preston	Westinghouse	Westinghouse	
1912	17,18	"	"	22/28	Brill	Siemens	Siemens	
1912	19-24	Single-deck	"	30	Preston	"	"	
1921	25-32	Enclosed	Brush	30/42	Brush bogies	Metro-Vickers	Metro-Vickers	

LANCASTER CORPORATION TRAMWAYS — FLEET LIST

Date	Fleet Nos	Type	Builder	Seats	Truck(s)	Motors	Controllers	Notes
1902	1-10	Open-top	Lancaster RC&W	18/23	Brill	Westinghouse	Westinghouse	
1905	11-12	"	Milnes Voss	"	Mountain & Gibson	"	"	

MORECAMBE TRAMWAYS COMPANY — FLEET LIST

Date	Fleet Nos	Type	Builder	Seats	Truck(s)	Motors	Controllers	Notes
1887	1, 2	Open-top	Lancaster RC&W	20/24				a
1887	3, 4	Toastrack	"	28				a
1888	5, 6	Open-top	"	20/24				a
1889	7	"	"	"				a
1897	8-11	"	"	"				a
1898	12-15	"	"	?	?			a/b
1901	16, 17	"	"	?	?			a/b
1911	1-3	Single-deck petrol	Leyland/UEC	35	UEC	Leyland		
1913	4	Open petrol	"	35	"	"		

Notes: **a)** Horse-drawn **b)** Second-hand — source unknown

BLACKBURN CORPORATION TRAMWAYS — FLEET LIST

Date	Fleet Nos	Type	Builder	Seats	Truck(s)	Motors	Controllers	Notes
1899	28-35	Open-top	Milnes	30/30	Brill bogies	Siemens	Siemens	
1901	36-75	”	”	32/41	Peckham bogies	General Electric	BTH	
1907	76-81	Single-deck	UEC	40	Brill bogies	”	”	
1908	82-87	”	”	”	”	”	”	
1908	88	Single-deck Cross-bench	BCT	46	”	”	”	

MORECAMBE CORPORATION TRAMWAYS — FLEET LIST

Date	Fleet Nos	Type	Builder	Seats	Truck(s)	Motors	Controllers	Notes
1909	1, 2	Open-top	Lancaster RC&W	20/24				a
1909	3, 4	Toastrack	”	28				a
1909	5, 6	Open-top	”	20/24				a
1909	7	”	”	”				a
1909	8-11	”	”	”				a
1909	12-14	”	?	?				a
1919	13	Double-deck	EE	18/22				
1919	14	Toastrack	”	32				
1922	15	”	”	”				
1922	16	Double-deck	”	18/22				

Notes: All horse drawn **a)** Acquired with Morecambe Tramways Co take-over (qv)

BET CO LTD AND BARROW-IN-FURNESS CORPORATION TRAMWAYS — FLEET LIST

Date	Fleet Nos	Type	Builder	Seats	Truck(s)	Motors	Controllers	Notes
1903	1-7	Open-top	Brush	22/26	Brush	Dick, Kerr	Dick, Kerr	
1903	8-12	Single-deck Combination	”	38	Brush bogies	”	”	
1905	13, 14	Demi-car	BEC	22	BEC	Brush	Raworth	
1905	15, 16	Open-top	Brush	28/32	Brush	”	Brush	
1911	17-20	”	”	96	Brill bogies	”	”	
1913	21, 22	Single-deck	”	40	Brush bogies	”	”	
1914	23, 24	”	”	”	”	”	”	
1915	25, 26	”	Midland	”	Brill bogies	General Elec	BTH	a
1917	27, 28	Open-top trailer	Brush	18/30	Brush			b
1920	1-4	Single-deck combination	ER&TCW	32	Brill	Dick, Kerr	Dick, Kerr	c
1920	29, 30	Single-deck	Milnes	28	”	General Elec	BTH	d
1920	31, 32	”	Brush	”	”	”	”	e
1920	33, 34	”	Sheffield Corp'n	”	”	”	”	f
1921	35, 36	”	Brush	32	Peckham	EE	EE	

Notes:
a) Built 1900 for Potteries Electric Traction Co
b) Motorised c1920 (with Dick, Kerr Motors and Controllers)
c) Built 1900 for Southport Corporation
d) Built 1899 for Sheffield Corporation
e) Built 1900 for Sheffield Corporation
f) Built 1901 for Sheffield Corporation

DARWEN CORPORATION TRAMWAYS — FLEET LIST

Date	Fleet Nos	Type	Builder	Seats	Truck(s)	Motors	Controllers	Notes
1900	1-5	Open-top	Milnes	30/42	Brill bogies	Westinghouse	Westinghouse	
1900	6-10	,,	,,	,,	,,	General Elec	BTH	
1901	11-14	,,	,,	60	Brill	,,		
1905	15	Demi-car	Milnes Voss	22	M&G	Westinghouse	Raworth-Westinghouse	
1906	16, 17	,,	,,	,,	,,	,,	,,	
1915	18, 19	Open-top	UEC	30/36	Peckham bogies	Dick, Kerr	Dick, Kerr	
1921	20-22	,,	EE	,,	EE 'Burnley Bogies'	,,	,,	
1924	16, 17	Enclosed	Brush	30/42	'Burnley Bogies'	BTH	BTH	
1925-29	3,5,7 8,15	,,	Darwen Corp'n	,,	,,	,,	,,	
1933	10	,,	,,	,,	,,	Metro-Vickers	,,	
1933	9	,,	?	?	?	?	?	a
1933	11	Balcony	?	?	?	?	?	b
1936	23, 24	Enclosed, centre entrance, streamlined	EE	24/32	EE bogies	EE	EE	

Notes:
a) Ex-Rawtenstall **b)** Ex-Rawtenstall on truck of former No 11

PRESTON CORPORATION TRAMWAYS — FLEET LIST

Date	Fleet Nos	Type	Builder	Seats	Truck(s)	Motors	Controllers	Notes
1904	1-26	Open-top	ER&TCW	22/26	Brill	Dick, Kerr	Dick, Kerr	
1904	27-30	,,	,,	30/38	Brill bogies	,,	,,	
1912	31-33	Single-deck	UEC	40	,,	,,	,,	
1914	34-39	Balcony	,,	22/30	Preston Flexible	,,	,,	
1918	40-45	Single-deck		28	Brill	,,	"	a
1920	46-48	,,	Brush	,,	,,	,,	,,	b
1928	30, 40, 42	Enclosed	Preston Corp'n		22/40	Preston Standard	EE	,,
1929	13, 18, 22	Balcony	EE	22/30	,,	,,	,,	c

Notes:
a) Ex-Sheffield; two built by Milnes, two built by SCT between 1901 and 1903, DK equipment fitted by Preston
b) Ex-Sheffield, built 1903 **c)** Ex-Lincoln Nos 9, 10 and 11

LYTHAM ST ANNES CORPORATION TRAMWAYS — FLEET LIST

Date	Fleet Nos	Type	Builder	Seats	Truck(s)	Motors	Controllers	Notes
1903	1-20	Open-top	BEC	22/32	BEC	General Electric	BTH	
1903	21-30	,,	,,	,,	,,	,,	,,	
1905	31-40	Open-top/ Cross-bench	Brush	34/34	Brush	,,	,,	
1924	41-50	Balcony	EE	23/38	Peckham	EE	EE	
1933	51-54	Single-deck	,,	36	,,	,,	,,	a
1933	55	Enclosed	Brush	32/44	Brush Bogies	Dick, Kerr	Dick, Kerr	b
1934	56	,,	Preston Corp'n	22/40	Preston	,,	,,	c

Notes: **a)** Built 1924 for Dearne District Light Railway. **b)** Built 1925 for Accrington Corporation (No 39).
c) Built 1928 as Preston Corporation No 42.

CITY OF CARLISLE ELECTRIC TRAMWAYS — FLEET LIST

Date	Fleet Nos	Type	Builder	Seats	Truck(s)	Motors	Controllers	Notes
1900	1-3	Single-deck	ER&TCW	22	Brill	Walker	Dick, Kerr	
1900	4-15	Open-top	,,	22/23	,,	,,	,,	
1912	1-8	Double-deck	UEC	22/28	,,	General Electric	BTH	
1912	9-12	Single-deck	,,	24	,,	,,	,,	
1920	13	Open-top	ER&TCW	22/26	,,	Dick, Kerr	Dick, Kerr	a
1923	15	,,	EE	,,	Westinghouse	Westinghouse		
1925	14	,,	,,	,,	Dick, Kerr	Dick, Kerr		

Notes: **a)** No 13 believed to be ex-Ilkeston, Derbyshire

BURNLEY CORPORATION TRAMWAYS — FLEET LIST

Date	Fleet Nos	Type	Builder	Seats	Truck(s)	Motors	Controllers	Notes
1901	1-24	Open-top	Milnes	32/39	Brill	General Electric	BTH	a
1903	25-38	,,	,,	,,	,,	,,	,,	
1903	39-46	Single-deck	ER&TCW	44	Brill bogies	Dick, Kerr	Dick, Kerr	
1907	47	,,	UEC	40	Simpson&Park	General Electric	BTH	
1909	48-52	Balcony	HN	32/39	Brill bogies	,,	,,	
1910	53, 54	Single-deck	UEC	44	,,	,,	,,	
1911	55-57	,,	,,	,,	,,	,,	,,	
1913	58-67	Balcony	,,	32/39	,,	,,	,,	
1921	68-72	Single-deck	EE	44	,,	,,	,,	b

Notes: **a)** No 10 renumbered 68 in 1926 **b)** No 68 renumbered 73 in 1926

COLNE AND TRAWDEN LIGHT RAILWAY & COLNE CORPORATION — FLEET LIST

Date	Fleet Nos	Type	Builder	Seats	Truck(s)	Motors	Controllers	Notes
1903	1-6	Open-top	Milnes	22/28	Milnes	General Electric	BTH	
1905	7-10	,,	Brush	22/26	Brush	,,	,,	
1906	11-12	,,	Milnes	22/28	M&G	,,	,,	
1914	13	Balcony	UEC	52	Preston	,,	,,	
1921	2, 3	,,	EE	68	EE Burnley bogies	,,	,,	
1926	14-16	Enclosed	Brush	52	Peckham	Metro-Vickers	Metro-Vickers	

NELSON CORPORATION TRAMWAYS — FLEET LIST

Date	Fleet Nos	Type	Builder	Seats	Truck(s)	Motors	Controllers	Notes
1902	1-3	Open-top	Brush	40	Brush	General Electric	BTH	
1903	4-6	,,	,,	,,	,,	,,	,,	
1903	7, 8	Single-deck combination	ER&TCW	38	Brill bogies	Dick, Kerr	Dick, Kerr	
1904	9	,,	,,	,,	,,	,,	,,	
1912	10, 11	Balcony	UEC	55	Preston	Siemens	Siemens	a
1916	1-6	,,	,,	,,	,,	Dick, Kerr	BTH	a
1925	7-9	,,	Brush	,,	Brush	Metro-Vickers	Metro-Vickers	a

Notes: **a)** 'Lowbridge' design

BLACKPOOL CORPORATION TRAMWAYS — FLEET LIST

Date	Fleet Nos	Type	Builder	Seats	Truck(s)	Motors	Controllers	Notes
1885	1, 2	Open-top	Starbuck	56	Trunnions	Elwell Parker	Holroyd Smith	a
1885	3, 4	,,	Lancaster RC&W	16/16	,,	,,	,,	a
1885	5, 6	,,	,,	22/22	,,	,,	,,	a
1885	7, 8	Open-top Cross-bench	Starbuck	48	,,	,,	,,	a

BLACKPOOL CORPORATION TRAMWAYS — FLEET LIST Continued.

Date	Fleet Nos	Type	Builder	Seats	Truck(s)	Motors	Controllers	Notes
1891	9, 10	Open-top	Milnes	56	Trunnion	ECC	ECC	a
1895	11, 12	"	Lancaster RC&W	36/44	Equal wheel bogies	"	"	a
1896	13, 14	"	"	"	"	"	"	a
1898	15, 16	Open-top 'Dreadnought'	Midland	36/50	"	Siemens	Siemens	a
1900	17-26	"	"	44/49	Midland bogies	BTH	BTH	
1901	27-41	Open-top	"	24/39	Midland	"	"	
1902	42-53	"	HN	34/41	HN bogies	"	"	
1902	54-61	"	Midland	44/49	Midland bogies	"	"	
1911	62-64	Balcony	UEC	28/38	Preston Flexible	"	"	
1911	65-67	"	"	28/36	Preston equal wheel bogies	"	"	
1911	69-70	Toastrack	"	69	"	"	"	
1912	68	Balcony	"	28/36	"	"	"	
1912	71-80	Toastrack	"	69	"	"	"	
1913	81-86	"	"	"	"	"	"	
1914	87-92	"	"	"	"	"	"	
1919	93-98	Open-top	Milnes	30/42	McGuire bogies	Westinghouse	Westinghouse	b
1920	126-135	Cross-bench	"	48	Milnes bogies	GE (USA)	GE (USA)	c
1920	136-138	"	"	"	"	Westinghouse	Westinghouse	d
1920	106-111	Single-deck	"	"	"	GE (USA)	GE (USA)	e
1920	101-105	"	"	"	"	"	"	f
1920	139-141	Cross-bench	Milnes	48	Milnes bogies	Dick, Kerr	Dick, Kerr	g
1920	116-122	Single-deck combination	ER&TCW	55	Brill bogies	"	"	h
1920	123-125	Cross-bench	UEC	64	UEC bogies	Westinghouse	Westinghouse	i
1920	112-115	Single-deck	UEC	48	"	GE (USA)	BTH	j
1923	99, 100	Balcony	BCT	32/46	HN	BTH	"	k
1923	33, 34	"	"	"	"	GE (USA)	"	k
1924	142-145	"	"	"	HN or EE bogies	BTH	"	k
1924	146-149	"	HN	"	"	"	"	
1925	150-152	"	"	"	"	"	"	
1925	36,38-41	"	BCT	"	"	"	"	k
1926	42, 49, 153-155	"	"	"	"	"	"	k
1927	28, 35, 37, 156-60	"	"	"	"	"	"	k
1927	161-166	Toastrack	"	64	"	"	"	k
1928	45, 47, 48, 50, 53	Balcony	"	32/46	"	"	"	k
1928	167-176	Single-deck	EE	48	"	GE (USA)	"	
1929	51, 177	Balcony	BCT	32/46	"	BTH	"	k
1933	200-224	Railcoach	EE	48	EE bogies	EE	EE	n
1934	225	Boat	"	56	"	"	"	o
1934	237-249	Open-top double-deck		40/54	"	"	"	p
1934	250-263	'Balloon'	"	40/44	"	"	"	p
1935	226-236	Boat	"	56	"	"	BTH	o
1935	264-283	Railcoach	"	48	"	"	EE	m/q
1937	284-303	"	Brush	"	EMB bogies	Crompton Parkinson	Allen-West	r
1939	10-21	Sun Saloon	EE	56	EE bogies	BTH	EE	
1952	304-328	Single-deck	Roberts	"	Maley & Taunton bogies	Crompton Parkinson	CP Vambac	s
1960	T1-10	Single-deck trailer	MCW	66	"	"	"	t
1972 onwards	1-13	Single-deck 'OMO'	BCT	48	EE bogies	EE	EE	u
1979	761	Double-deck 'OMO'	"	42/56	"	"	Westinghouse	v
1982	762	"	"	34/56	"	"	Brush	w

BLACKPOOL CORPORATION TRAMWAYS — FLEET LIST Continued.

Date	Fleet Nos	Type	Builder	Seats	Truck(s)	Motors	Controllers	Notes
1984	641-648	Centenary Single-deck	East Lancashire	52	BCT bogies	,,	,,	

ILLUMINATED TRAMCARS

Year	No	Car	From	Seats	Withdrawn
1925	-	Gondola	28 of 1901	20	1962
1926	-	Lifeboat	40 of 1901	20	1961
1937	-	Progress	141 of 1898	-	1958
1959	731	Blackpool Belle	163 of 1927	32	1978
1959	158	Double-deck	158 of 1927	78	1966
1959	159	Double-deck	159 of 1927	78	1966
1961	732	Rocket	168 of 1928	47	In use
1962	733	Wild West Locomotive	209 of 1934	35	In use
1962	734	Wild West Carriage	174 of 1928	60	In use
1963	735	Hovertram	222 of 1934	99	In use
1965	736	HMS Blackpool	170 of 1928	66	In use

NOTES:
a) Ex-Blackpool Electric Tramways Co conduit cars in 1892. All converted to overhead
b) Ex-London United, built 1901
c) Ex-B&F Nos 1-10 built 1898
d) Ex-B&F Nos 11-13 built 1898 (as trailers)
e) Ex-B&F Nos 14-19 built 1898
f) Ex-B&F Nos 20-24 built 1898
g) Ex-B&F Nos 25-27 built 1899
h) Ex-B&F Nos 28-34 built 1899
i) Ex-B&F Nos 34-37 built 1910
j) Ex-B&F Nos 38-41 built 1914
k) Built by Blackpool Corporation
l) Roofed during World War 2.
m) Nos 272-81 rebuilt as towing cars for twin-sets in 1958-62 and renumbered in 671-80 series in 1968
n) Survivors renumbered in 608-10 series in 1968
o) Survivors renumbered in 600-07 series in 1968
p) Class renumbered in 700-26 series in 1968
q) Class (except note m) renumbered in 611-20 series in 1968
r) Survivors renumbered in 621-38 series in 1968
s) Survivors renumbered in 641-64 series in 1968
t) Class renumbered to 681-90 series in 1968
u) Rebuilt from English Electric Railcoaches
v) Rebuilt from 'Balloon' No 725 (ex-No 262)
w) Rebuilt from 'Balloon' No 714 (ex-No 251)

Colne Corporation were very proud of their trio of fully-enclosed Brush cars of 1926, as they were reckoned to be the fastest in the district. No 16 was photographed adjacent to the depot at Heifer Lane, sporting the maroon and cream livery which superseded the royal blue and white, inherited from the company in 1922 or thereabouts. To clear the station bridge, covered-top cars were built to a 15ft 7in height. A corporation normal control Guy bus is in the background. *Neville Lockwood Collection*

APPENDIX 'A'

SUMMARY OF PRINCIPAL SYSTEMS
(ELECTRIC UNLESS SHOWN OTHERWISE)

SYSTEM	OPENED	CLOSED	GAUGE	MAX LENGTH (MILES)	LIVERY	ANTECEDENT COMPANIES	OPENED	NOTES
Accrington	2.8.07	6.1.32	4ft 0in	9.92	Red/Cream	Accrington Corporation Steam Tramways Co (S)	5.4.86	a/g
Barrow	1.1.20	5.4.32	4ft 0in	6.39	Green/Cream	Barrow-in-Furness Tramways Co (S) BET Co Ltd	11.7.85 6.2.04	
Blackburn	20.3.99	3.9.49	4ft 0in	14.73	Green/Cream	Blackburn Corporation Tramways Co (S/H) Blackburn & Over Darwen Tramways Co (S)	28.5.87 16.4.81	g
Blackpool	11.9.92		4ft 8½in	11.95	Green, Red/White, Green/Cream	Blackpool Electric Tramways Co (C) Blackpool & Fleetwood Tramroad Co.	29.9.85 14.7.98	g g
Burnley	16.12.01	7.5.35	4ft 0in	13.05	Chocolate/Primrose	Burnley & District Tramways Co (S)	17.9.81	b
Carlisle	30.6.1900	21.11.31	3ft 6in	5.73	Chocolate/Cream, Green/Cream	City of Carlisle Electric Tramways Co.	30.6.1900	c
Colne	25.3.14	6.1.34	4ft 0in	5.23	Blue/White/Maroon/Cream	Colne & Trawden Light Railway Co.	30.11.03	b/g
Darwen	17.10.1900	5.10.46	4ft 0in	4.36	Vermilion/Purple, Vermilion/Cream	Blackburn & Over Darwen Tramways Co (S)	16.4.81	g
Lancaster & District (H)	2.8.90	21.12.21	4ft 8½in	4.30	Cream			
Lancaster	14.1.03	31.3.30	4ft 8½in	2.99	Chocolate/Primrose			
Lytham St Annes	28.10.20	28.4.37	4ft 8½in	6.31	Blue/Cream	British Gas Traction Co (G) Blackpool, St Annes & Lytham Tramways Co (G/E)	11.7.96 1898	g
Morecambe (H)	2.7.09	6.10.26	4ft 8½in	2.38	Green/Cream	Morecambe Tramways Co (H)	3.6.87	
Morecambe Tramways Co (P)	2.7.09	24.10.24	4ft 8½in	1.19		Morecambe Tramways Co (H)	3.6.87	d
Nelson	23.2.03	6.1.34	4ft 0in	2.75	Red/White/Brown/Cream			b/g
Preston	7.6.04	15.12.35	4ft 8½in	10.53	Maroon/Cream	Preston Tramways Co (H) Preston Corporation Tramways Co (W Harding) (H)	20.3.79 14.4.82	e
Rawtenstall	15.5.09	13.3.32	4ft 0in	11.75	Maroon/Cream	Rossendale Valley Tramways Co	31.1.89	f

KEY:
C — Conduit electric E — Electric overhead
G — Gas   H — Horse
P — Petrol   S — Steam

NOTES
a) Operated 2.9 miles on behalf of Haslingden Corporation. b) Constituent of Burnley, Colne and Nelson Joint Transport, 1.4.33.
c) Renamed City of Carlisle Electric Transport Company, 3/26. d) Remainder of system after Morecambe Corporation take-over.
e) 3ft 6in gauge. f) Operated 2.36 miles on behalf of Bacup Corporation.
g) Running powers over other systems.

126

APPENDIX 'B'
NOTABLE DATES

1870   Tramways Act.
1879   Preston Tramways Co commences.
1881   Blackburn & Over Darwen Tramways Co commences.
        Burnley & District Tramways Co commences.
1882   Preston Corporation Tramways Co commences.
1885   Barrow-in-Furness Tramways Co commences.
        Blackpool Electric Tramways Co commences.
1886   Accrington Corporation Steam Tramways Co commences.
1887   Blackburn Corporation Tramways Co commences.
        Morecambe Tramways Co commences.
1889   Rossendale Valley Tramways Co commences.
1890   Lancaster & District Tramways Co commences.
1892   Blackpool Corporation takes over tramway company.
1896   British Gas Traction Co commence operations in St Annes and Lytham.
1898   Blackpool & Fleetwood Tramroad Co commences.
1899   Blackburn Corporation Tramways take over company lines.
1900   City of Carlisle Electric Tramways Co commences.
        Darwen Corporation Tramways take over company lines.
1901   Burnley Corporation Tramways take over company lines.
        Lancaster Corporation Tramways commence.
1903   Nelson Corporation Tramways commence.
        Colne & Trawden Light Railway commences.
1904   Preston Corporation Tramways take over company lines.
        Barrow Tramways Co acquired by BET Co Ltd.
1907   Accrington Corporation Tramways take over company lines.
1909   Rawtenstall Corporation Tramways take over company lines.
        Morecambe Corporation Tramways take over part of company lines, remainder stays with company.
1914   Colne Corporation Light Railways take over company lines.
1920   Barrow Corporation Tramways take over company lines.
        St Annes Corporation Tramways take over company lines.
1921   Lancaster & District closes.
1924   Morecambe Tramways Co closes.
1926   Morecambe Corporation Tramways close.
1930   Lancaster Corporation Tramways close.
1931   Carlisle Tramways close.
1932   Accrington Corporation Tramways close.
        Rawtenstall Corporation Tramways close.
        Barrow Corporation Tramways close.
1933   Burnley, Colne & Nelson Jt Cttee formed.
1934   Former Nelson system closes.
        Former Colne system closes.
1935   Former Burnley system closes.
        Preston Corporation Tramways close.
1937   Lytham St Annes Corporation Tramways close.
1946   Darwen Corporation Tramways close.
1949   Blackburn Corporation Tramways close.

In 1921, a trio of all-English Electric cars was delivered to Darwen Corporation, the final one, No 22, being depicted here. They set a precedent for future new acquisitions in that 'Burnley Bogies' were chosen, a trend to be broken only by the last pair of new cars 15 years later. *Roy Marshall Collection*

BIBLIOGRAPHY

W. H. Bett and J. C. Gillham — *The Tramways of North Lancashire* — Light Rail Transit Association.

Gordon Biddle — *The Railways Around Preston* (Canal Tramway) — Foxline, 1989.

Alan Catlow — *Burnley, Colne and Nelson Joint Transport* — Wyvern Publications, 1985.

Richard Catlow and Tom Collinge — *Over the Setts* — Countryside Publications, 1978.

Ian L. Cormack — *Seventy-Five Years on Wheels* (Barrow) — Scottish Tramway Museum Society, 1961.

Dorothy Harrison — *The History of Colne* — Pendle Heritage Centre, 1988.

George S. Hearse — *Tramways of the City of Carlisle* — Author, 1962.

Geoffrey W. Heywood — *The Tramways of Preston* — Tramway Review 67-71, 1971-72.

W. G. S. Hyde and F. K. Pearson — *The Dick, Kerr Album* — Authors, 1972.

J. Joyce — *The Story of Passenger Transport in Britain* — Ian Allan, 1967.

J. Joyce, J. S. King and A. G. Newman — *British Trolleybus Systems* (Ramsbottom) — Ian Allan, 1986.

Steve Palmer and Brian Turner — *Blackpool by Tram* — Transport Publishing Co, 1978.

Steve Palmer — *Blackpool and Fleetwood by Tram* — Platform 5 Publishing, 1988.

Donald F. Phillips — *The Tramways of Lytham St Annes* — Tramway Review, 14, 1954.

Robert W. Rush — *The Tramways of Accrington 1886-1932* — Light Railway Transport League, 1961.

S. Shuttleworth — *The Lancaster and Morecambe Tramways* — Oakwood Press, 1976.

David St John Thomas — *The Country Railway* (Legislation) — David and Charles, 1976.

Brian Turner — *Blackpool to Fleetwood* — Light Railway Transport League.

ACKNOWLEDGEMENTS

Enthusiastic encouragement was readily forthcoming during the preparation of this book, as was the invaluable assistance from numerous sources, and the author is indebted to them all. Help in providing photographs was especially appreciated, and it is hoped that accurate accreditation has been achieved; in respect of older prints, this is often difficult, original copyright sometimes being lost in the 'mists of time'. A number of organisations have been particularly helpful: Leicestershire Museums and Lancashire and Cumbria Libraries, to name but three. Some individuals warrant special mention: Rosie Thacker and Glynn Wilton at the National Tramway Museum, Peter Iddon at Blackburn Transport, Barry McLaughlin at the *West Lancashire Evening Gazette*, Alan Catlow and Neville Lockwood of the BCN Society, Roy Marshall, A. D. Packer and, of course, the author's wife, Vivienne, for her tolerance with an inevitable increase in his usual obsession with transport matters! Finally, a special thank you must go to Marjorie Kelly, whose nimble fingers at the keyboard and patience with scrawled drafts proved indispensable.

Left: In 1912, 12 of the original fleet were replaced by eight double and four single-deck cars, No 2 of the former being seen here towards the end of the system's life. It was a standard Dick, Kerr product. *National Tramway Museum*

Front cover: Now preserved Blackpool 'Standard' No 40 is seen in its home town on 28 October 1962. Prominent in the background is the famous tower which celebrates its centenary in 1995. *Geoff Lumb.*